Heather Wilson

Get Back Up

Dave Gallagher

Get Back Up

Lessons in Servant Leadership

BY DR. HEATHER WILSON
AND GENERAL (RET.) DAVE GOLDFEIN

UTEP Press
The University of Texas at El Paso

Published by UTEP Press, an imprint of The University of Texas at El Paso.

Lucas Roebuck, Publisher

Jenn Crawford, Managing Editor

Danielle Waters, Cover Design

UTEP Press
500 W. University Ave.
El Paso, TX 79968
utep.edu/press

For Ava, Miles, Miriam, Piper, Rae, Remi, Ruby, George, and Albert and more grandchildren if we are blessed to have them. May they be inspired to pursue lives as servant leaders.

Reflections on
Get Back Up: Lessons in Servant Leadership

In Get Back Up: Lessons in Servant Leadership, *Dave Goldfein and Heather Wilson recount the moments that have shaped, challenged, and inspired them. Each story, whether joyful, bittersweet, or sober, offers a candid look into what servant leadership can and should look like. Dave and Heather offer timeless lessons that everyone can use. A must-read for anyone who believes that leadership starts with serving others.*

—Steve Schwarzman, Chairman, CEO, and Cofounder of Blackstone

Dr. Wilson and General Goldfein are meaningful reminders of the value of and need for servant leaders. They approach this book as they have approached their lives—with humility, foresight, selfless sharing of themselves, and a commitment to improving our nation. While their book should be required reading for every CEO and leader, its greatest impact may well be on the next generation. Whether a cadet, college student, or young professional, the life and leadership lessons they share will entertain, educate, and inspire.

—Montana Governor Steve Bullock (2013–2021)

A uniquely enriching guide for both the aspiring young leader and the experienced leader devoted

to upping their game, this jewel of a book is packed with powerful lessons-learned and distilled insights. Provided by two servant leaders whose character met every challenge from the halls of Congress to the battlefields of war and beyond, their humility, honesty, and raw competence shine through on every page. Dr. Wilson and General Goldfein's book lights the path for leaders who choose to serve others.

—General Jim Mattis, United States Marines, Retired, and 26th Secretary of Defense

In aviation, as in leadership, turbulence is inevitable. Get Back Up *shows how two extraordinary leaders, Dave Goldfein and Heather Wilson, navigated adversity with humility and courage. This volume offers young leaders practical insights from lives shaped by principle, service, and the courage to step forward when it matters most.*

—Ambassador Barbara Barrett, 25th Secretary of the Air Force

In this powerful and deeply personal book, Heather Wilson and Dave Goldfein deliver timeless lessons in servant leadership—grounded in humility, service before self, and a deep commitment to others. These are not abstract ideals; they're principles both authors have lived. As the First Chief of Space Operations, I benefited greatly from Dave's leadership and partnership. His steady, values-driven leadership was instrumen-

tal in helping establish the foundation of our new United States Space Force.

—General John W. "Jay" Raymond, United States Space Force, Retired, 1st Chief of Space Operations

In Get Back Up, *Dave Goldfein and Heather Wilson remind us of a crucial paradox: that power is best expressed through humility, credibility, and approachability. The most important lesson the reader can learn is the timeless concept of servant leadership. This is demonstrated through their love for the Department of the Air Force; it was deeply personal. Their teamwork showed in their love and passion for our Airmen, Guardians, their families, and ultimately the institution. They truly demonstrated on a daily basis that they were a part of something bigger than themselves.*

—General Lori Robinson, United States Air Force, Retired, Former Commander of US Northern Command and North American Air Defense Command, 2016–2018

As a young woman who grew up under the shadow of war in Afghanistan and chose to serve my country, I see in this book a rare kind of strength—quiet, steady, and rooted in purpose. . . . Their words will stay with you long after the last page.

—Niloofar Rahmani, author of *Open Skies: My Life as Afghanistan's First Female Fixed-Wing Air Force Pilot*

Contents

About the Authors

Dave Goldfein and Heather Wilson started their careers on the same day and in the same place: the United States Air Force Academy in Colorado Springs, Colorado. Their careers took different paths but converged almost forty years later at the Pentagon in Washington, DC.

Dave became a fighter pilot and military commander, was shot down by a surface-to-air missile over Serbia, and spent two years as the Commander of Coalition Air and Space Forces in Iraq and Afghanistan before becoming the Director of the Joint Staff in the Pentagon.

Heather earned her PhD as a Rhodes Scholar, worked in Europe during the Cold War, and served on the National Security Council Staff for President George H.W. Bush. She was a cabinet secretary in state government, elected and reelected to the United States Congress for ten years, and then became a university president.

Heather and Dave were brought back together in May 2017 as Secretary and Chief of Staff of the Air Force under Secretary of Defense James Mattis. During their service together, they talked a lot about leadership. Many of the featured stories originate from the time they spent working together.

Dave is now a senior advisor to Blackstone, the world's largest alternative asset manager, and is chairman of the board of Google Public Sector and the USO.

Heather is the president of The University of Texas at El Paso.

Introduction
Where It Started

Heather

It was 1976, and I was a junior in high school growing up in rural New Hampshire. The black-and-white television in my mom's bedroom reported on the evening news that the first class to include women had entered the United States Air Force Academy.

That was interesting.

Even at 15 years old, I was a competitive and adventurous kid, prone to do things a bit out of the ordinary. That image of women entering that beautiful campus in the Rocky Mountains stuck with me. A few days later, I went to talk to my grandfather about it. He had been one of the first pilots in the United Kingdom's Royal Air Force in World War I and came to America in 1922 in search of work. A barnstormer and mechanic, George Gordon "Scotty" Wilson opened little airports in New England in the 1920s and 1930s. In World War II, he was part of a group of civil aviators who patrolled for submarines along the Atlantic coast, towed targets for gunnery practice, and ferried parts to airfields to support the military. That group eventually became known as the Civil Air Patrol, the official auxiliary of the US Air Force.

I deeply admired my grandfather. As his only granddaughter, I was something special to him. I knew he wanted the best for me, and he was proud of how well I was doing in school, but I was also a little afraid of what he might say. It was, after all, 1976. Opportunities were opening for women, but thinking about going to the Air Force Academy was more than a bit unusual.

He was sitting in his customary armchair in the living room of my grandparents' small house when I told him I was thinking about applying to the Air Force Academy. There was a long pause as he stared at his aged hands resting in his lap. Then, in his soft Scottish brogue, he said, "Well, I flew with some women in World War II—towing targets and ferrying airplanes. The WASPS. They were pretty good 'sticks.' So, I guess that would be okay."

With his blessing, I applied to the Air Force Academy.

On Saint Patrick's Day in 1978, my congressman's office told me that I had been accepted. It was a full ride scholarship and an opportunity to chart my own course.

A few days after graduating from high school that same year, I went to see my grandfather to say goodbye before leaving to attend the Academy. I was 17, the same age he had been when he lied about his age and joined the RAF. The same age his son, my father, was when he enlisted in the Army Air Corps just after the end of World War II.

My father, also a pilot and mechanic, died in a car accident when I was six years old. As a child, I don't think I really understood how turning the arc of my life toward aviation might have felt to my grandfather. Now with children and grandchildren of my own, I have a deeper understanding of what my decision meant to him.

The day I left for the Air Force Academy was the only time in my life I saw my grandfather cry. A single tear traced the wrinkles of his aged face. He didn't acknowledge it was there and neither did I. I wonder now what he was thinking about—about the son he lost too soon, about the life he had lived in aviation, about the granddaughter he loved and who loved him? I didn't ask. In that moment of emotional connection, it was likely some of all those things.

My brother drove me in his pickup truck to the airport in Hartford, Connecticut. I had a small brown suitcase with a single change of clothes, a dozen sets of white underwear, and combat boots we'd been advised to purchase early to break in. I had a 35 mm camera that was my high school graduation present from my mother. Most important of all, I had a one-way ticket to Colorado Springs, Colorado.

I was on my way. I didn't know it then, but it would be the adventure of a lifetime.

Dave

I was in high school at Ramstein Air Base, where my father worked as a colonel on the NATO staff. My older brother had been at the Air Force Academy for three years and was at the top of his class as a leader, eventually rising to wing commander—the highest-ranking cadet.

I applied to the Academy but received a polite rejection letter rather early in the process with every box checked as to the reasons why: academics, athletics, leadership, extracurricular activities, community involvement, the list goes on.

So, I applied to the University of Wyoming where my best friend, Bob Ihle, planned to attend. My passions as a teenager were scouting, the outdoors, and music. Where better to study forestry and pursue my dream of becoming a guitar strumming, back-country ranger at a national park? They let me in.

Then I got the phone call that would change my life.

A lieutenant colonel from the Academy admissions office called me at the base commissary where I was bagging groceries for tips. "We haven't heard back from you on the preparatory school scholarship we offered," he said. "Are you going to take it?" The thing was, I had no idea what he was talking about. As it turns out, they sent an offer letter to my aunt in New York, but she never forwarded it to us in Germany.

I thanked him and kindly informed him I was headed to Wyoming to study forestry and to please give the scholarship to someone else. There was a long silence on the phone before he asked if I wanted to discuss it with my parents first. Meanwhile, my boss was yelling for a bagger. I quickly told him I would, though it wouldn't make any difference, because I knew I was set on Wyoming, and my first tuition check had already been paid. After all, Bob would leave in a week, and we'd already discussed him setting up our dorm room. I wasn't far behind.

I went home for supper.

My dad was reading the *Stars and Stripes* newspaper over a plate of lasagna when I casually mentioned the phone call and that I had turned the offer down since everything was in motion for Wyoming.

There was a long pause.

My dad slowly folded the paper and set it aside. His steely blue fighter pilot eyes that had seen combat over the skies of Vietnam locked onto mine. I remember my younger brother, Mike, also looking at me like I was a bit crazy.

"Let me make sure I understand this," my father said deliberately. "Someone from the Air Force Academy called you today and offered you money to go to school.

And you told him no, that you would rather use my money to go to school."

Let's just say I called Lieutenant Colonel Jackson back. He didn't seem surprised and hadn't given the scholarship away.

A year later, after time at preparatory school, I received a different kind of letter from the Air Force Academy telling me that I had been accepted in the class of 1982.

I'm forever grateful to that lieutenant colonel.

Chapter 1

Am I Worthy?

Dave

On a full moon night, while flying a combat mission over Serbia on May 2, 1999, I became a pilot with more takeoffs than landings. Once I pulled the ejection handle and rode the rocket seat out of my dying F-16, my life depended on the courage and commitment of a small team of special operators who risked everything to bring me home.

You can read more about the details of this experience and the leadership lesson it taught me in the next chapter, but besides "get back up," this experience taught me another equally important lesson.

During the final moments of the rescue, as the sun began to rise and as the team fought their way to me, overcoming surface-to-air missiles that enemy troops were firing from below, I thought about these brave young warriors and how lucky I was to be from a nation that wouldn't leave anyone behind. America was not going to rest that night until I was safely back in the arms of my wife and daughters.

Was I worthy of their risk?

This question—am I worthy?—thus became my daily "mirror check" in every position I was privileged to fill. It remains my mirror check to this day and will be for the rest of my leadership journey.

Leadership is a precious gift offered by those entrusted to our care. As servant leaders, we must work every day to earn and re-earn this gift. The way we do this is through how we act when nobody is watching, how we live our lives, how we approach the tough decisions when good options are long gone, and how we treat and take care of those who choose to follow us.

As soon as we start feeling entitled to either the position or the perks of rank and responsibility, we begin to deviate from true servant leadership. Trust grows over weeks, months, and years, but it can be lost in a moment of indiscretion or in a sense of entitlement.

To be truly worthy of the gift of leadership, we must also understand the difference between character and reputation. Character defines who we are and forms the very essence of a servant leader. Reputation is how others see us after watching our performance over time. If we focus on the former, then the latter will take care of itself. (The same is not always true in reverse.)

Knowing the importance of character in leadership makes us reflect on our own actions and values. In doing so, we are challenged to evaluate our worthiness through the lens of courage, humility, and service.

Am I worthy of the risk those young warriors took to bring me home those many years ago?

Am I worthy of the trust extended to me by the parents of the young men and women who chose to join the Air Force—those young people who are the greatest treasure in our nation's arsenal? Am I worthy of leading those who look to me to make tough decisions with character, courage, and competence?

Am I worthy of standing before my grandchildren, to whom I dedicate this book, in the hope that they, too, will be inspired to pursue a path of servant leadership?

The only answer I have found appropriate over the years is one that brings me to prayer. Please God, I hope so. I hope I am worthy.

Heather

It was 2007. I was called into a briefing in the windowless hearing room of the House Permanent Select Committee on Intelligence, which, at the time, was high in the Capitol building underneath the cast iron dome.

I had been working for over a year on oversight hearings and draft legislation to update our foreign intelligence surveillance laws, which failed to keep up with advancements in technology and the changing nature of global threats. We came close to an update in late 2006 but ultimately didn't get new legislation across the finish line.

Our government isn't set up to be efficient; it is set up to protect us from tyranny. Even easy, commonsense things are hard to get through the Congress. And updating intelligence collection laws—balancing national security with protection of civil liberties—in a technical area of the law wasn't easy. But it was important. The 2006 draft legislation died, as all legislation does, when the new Congress was sworn in. And with a change in power in the House of Representatives, I was now in the minority. I wasn't setting the agenda, but in that committee room, all of us knew there were problems with the wiretapping laws and all of us knew that they needed to be fixed.

The law on intelligence collection, written in 1978, was specific to the technology of its time. In those days, almost all local telephone calls were made on phones attached to wires—landlines—and almost all international calls were transmitted over the air, typically using radio waves.

By 2007, however, the technology had completely reversed. Nearly all local calls were on cell phones using wireless communication networks, and a lot of international calls were routed by computers on fiber-optic cables laid on the ocean floor. According to the original text of the law, any communications sent over the air could be collected, but if intelligence agencies wanted to touch a wire in the United States—even to gather the communications of terrorists overseas—then they needed a warrant from a judge. That made no sense, and the law needed updating.

That day, underneath the Capitol dome, we were briefed on a particular case.

The Islamic State insurgency in Iraq was still threatening Americans, and three Soldiers from the 10th Mountain Division had been captured by Islamic extremists in a dangerous zone in Iraq known as the "triangle of death." While the American military immediately began searching house-to-house to find the Soldiers, the Department of Justice went to the Foreign Intelligence Surveillance Court in Washington, DC, for permission to collect communications on the group that the Army thought was responsible.

Think about that for a second.

To get information about the communications and whereabouts of the terrorists holding our servicemen hostage in a combat zone—terrorists who, mind you, have no rights under the US Constitution and have never stepped foot in America—the US Army was sending lawyers to go to a court in Washington, DC.

The briefer, a lawyer from the Justice Department, explained to Committee members how fast they had responded. Within twenty-four hours, a petition was written, and a warrant was approved by a judge in the middle of the night. My colleagues asked more questions about the missing Soldiers and what was being done to find them.

In congressional hearings, questioning alternates between Republicans and Democrats and proceeds in order of seniority. I had one question when my turn came.

"You say it took you only twenty-four hours to get permission to look at the communications of a terrorist group who might be holding Americans in a combat zone. If it was your son who was being held hostage, is that fast enough?"

The lawyer, who until that moment had vigorously defended the speed of their legal work, paused and looked down at the witness table. Then he looked up at me with sadness.

"No, ma'am. It's not."

When it was over, I left the hearing and walked in silence through the tunnels that connected the Capitol to the Cannon House office building. I stepped through my office door, flanked on one side by the yellow and red of the New Mexico flag and, on the other, the Stars and Stripes. I went into the small restroom in the corner of my personal office, shut the door, and looked at myself in the mirror, thinking about the laws we had not yet managed to change. I gripped the edge of the white sink with both hands, bowed my head, and wept.

We had failed them. I had failed them.

Article I of the Constitution grants the Congress the power to "raise and support" armies. Not the President, not the judiciary, but the elected representatives of the people. My constituents sent me to Washington to represent them, and, as a member of Congress, I felt I had not done my job well enough. We hadn't fixed this law.

As a servant leader, the "mirror check" means asking yourself, "Am I worthy?" This can be especially painful on days when you're not living up to what people have a right to expect of you.

Am I worthy of their trust?

Am I worthy of their hard-earned tax dollars?

Am I doing the best I can for the people who gave me the privilege to serve?

Am I worthy of their sacrifice?

As a servant leader, to be worthy of those we serve, our job is to do the best we can with the gifts we've been given. This is the task, but it is one of the hardest tests of leadership.

A little over a year later, after hundreds of hours of work and negotiation, I was in the Rose Garden at the White House when President George W. Bush signed the Foreign Intelligence Surveillance Act Amendments that updated our laws. To tell you the truth, I don't remember anything President Bush said that day. My mind was someplace else. I was thinking about three Soldiers from the 10th Mountain Division.

Heather and Dave

Leadership demands that we ask hard questions . . . starting with ourselves. Only by doing so can we truly understand that leadership is as much a gift as it is a burden, and that it is as much an opportunity as it is a challenge. As leaders, it is up to us to ensure that we never lose sight of the sacrifices made by others who have enabled and entrusted us to lead. Stories of those who have laid down their own interests and, for some, their lives, for a cause greater than themselves should set a high bar for what it means to be worthy of leadership. Such narratives are not merely tales of valor; they are the benchmarks against which we must measure our own commitment to service. The question "Am I worthy?" grounds us in our role and holds us accountable by demanding honesty and humility. It reminds us of the reason behind our positions of service and the people we are privileged to lead.

Every day, check yourself in the mirror
and ask: Am I worthy?

Chapter 2

Get Back Up

Dave

It was 1999 and 1:45 in the morning when my wife, Dawn, woke with a start. She looked at the clock on the nightstand wondering if something was amiss in our home just outside Aviano Air Base in northern Italy.

Dawn got up and walked around the house with a bad feeling and a baseball bat wondering if there was an intruder. She knew there was fierce local opposition to the war and was told to be wary of protestors. She needed to know all was safe for her and our oldest, Danielle, who had crawled into bed with her at some point in the night. Our youngest, Diana, was at a friend's house for a sleepover. Satisfied the house was safe, she went back to bed but laid awake until sunrise, unable to sleep.

We'd been stationed there for two years with our 12- and 9-year-old daughters. I was out flying a combat mission that night over Serbia as part of Operation Allied Force, which was the NATO effort to stop Serbia from ethnic cleansing in Kosovo. NATO leadership had decided to engage in the first combat operations since the alliance was formed after World War II. An ethnic cleansing campaign on its eastern border was unacceptable, and an air campaign was deemed the most appropriate response.

Flying in combat from home was a new experience for us. Mow the grass in the day then go get shot at that night. There was no parallel in recent history. Fighter pilots had always deployed forward without their families during previous conflicts. It certainly was Dawn's and my experience ten years earlier when I deployed to fly in Operation Desert Storm to drive Iraqi President Saddam Hussein from Kuwait.

This time was different. I was now commander of the 555th "Triple Nickel" fighter squadron leading missions in an F-16 Fighting Falcon, including the opening strike of the war on March 24. Slobodan Milosevic, president of Serbia, had made the wrong decision following a meeting with Ambassador Richard Holbrooke and Lieutenant General Mike Short in Belgrade weeks earlier.

Mike told me what he told Milosevic: "As the military leader responsible for the air operation, I have two options for you to consider. In one hand I have a U-2—a high-altitude reconnaissance aircraft with no munitions. In the other hand I have a B-2—a bomber that will result in a devastating outcome for your country if I am forced to use it. This is your choice, Mr. President. I pray you choose wisely."

He didn't.

So, we were now at war and the Triple Nickel was leading the effort along with our sister squadron, the 510th Buzzards.

Our mission on May 2, 1999, was to search out and destroy enemy missile sites in Serbia that had been shooting at NATO aircraft every night for the previous six weeks. A little over a month earlier, they successfully downed an F-117 fighter near Belgrade. The pilot, Lieutenant Colonel Dale Zelko, was successfully rescued by NATO forces. Our job that night was to make clear to the Serbians that there was a cost for shooting at us.

For the first month of the war, I led every mission as commander. I bore in mind something my boss did during Desert Storm that made a significant impact on me and my fellow fighter pilots.

Lieutenant Colonel Billy Diehl was a Vietnam veteran with a MiG kill on his record. He was fearless under fire, and we all were grateful to have a gifted combat veteran in the lead. About a month into the air campaign, Billy started flying on each of our wings. He told us, "There will be lots of medals and recognition when this is all over. But there is one thing I am going to do for you that my commander did for me in Vietnam. I am going to fly on each of your wings on a combat mission and give you a check ride. This will document in your official flying record that you led a combat mission over Iraq in a way no medal citation will represent."

It was now my chance to do the same for my squadron.

So, on May 2, I was flying as Hammer 4 on the wing of my chief of plans, Captain Adam "Jammer" Kavlick. It was a check ride neither of us would forget.

After taking gas from a refueling tanker over Bosnia, we headed toward our targets in Serbia. At approximately 1:45 a.m., we saw missiles being launched at us from sites we knew about. Staying outside lethal range, our job was to deliver our precision-guided weapons on the launch sites. We have a saying in the fighter aviation business: "Better to go after the archer than the arrows."

We were after the archers.

As is usual for many shootdowns, the missile we didn't see coming is the one that hit me. Serbian Lieutenant Colonel Zoltán Dani positioned his missile battery directly underneath our flight path and shot at me from below. Others in the formation who witnessed the explosion said it took about seven seconds from launch to impact.

We later nicknamed Zoltán "MacGyver" after the popular television series with actor Richard Dean Anderson who specialized in rewiring technology to achieve great results. Zoltán had rewired his kit so we would not find it during the mission, just as he had during his successful shootdown of Dale's F-117. Every service has a MacGyver. It was unfortunate for me and Dale that Zoltán happened to be in the air defense business.

Speaking of bad luck: A few weeks earlier on a training mission, my F-16 had been struck by lightning during recovery to the base. There was a loud explosion followed by a burning smell in the cockpit. Thankfully, the lightning only hit my wingtip and exited the tail, causing minimal damage. I was able to fly safely back to base with my wingman.

When the missile hit the back of my aircraft, it felt the same as a lightning strike. A loud explosion. An acrid smell in the cockpit. Connecting these dots, just as I had done weeks earlier, I turned the jet toward safety and made a radio call to my formation to let them know I had been hit.

Strange how life prepares you in ways you didn't appreciate the first time around.

But unlike before, the smoke in the cockpit got worse. The shaking made instruments impossible to read. It didn't take me long to determine that the missile had taken out my engine. Not a good thing for a single-engine fighter. I told the formation I was trying to restart it while searching for a good place to jump out if needed.

After several unsuccessful attempts, it was clear I was piloting what had become a very expensive glider. The $35 million aircraft was going to hit the ground. The only question was whether I was going to be in it when it did so.

I made a radio call to my formation. "Start finding me, boys."

I knew their radar information might be the best data available for the rescue team to find me. I also directed everyone to stop discussing my location, expecting that the enemy was listening. The race was on to get to me as soon as I pulled the handle and ejected.

Time slowed as I prepared to eject. I remember taking off my night vision goggles and kneeboards before assuming the correct position with head back against the seat and feet planted firmly on the floor. Funny what goes through your mind at a moment like this. I had borrowed a new kneeboard that night from one of my squadron mates, Lieutenant Colonel "Conan" Coan, to try it out. As I placed it to the side, I remember thinking to myself, Sorry, Conan.

I took one more look outside to make sure the jet would not crash in a populated area and pulled the handle between my knees. For what seemed like a long time, nothing happened. Oh crap, I thought. The seat is jammed. Then the canopy blew off in a loud explosion and up into the night sky I went as the rockets fired in the ACES II ejection seat.

The quiet from under the parachute was intense. I had no idea how loud it had become inside my dying aircraft, which was the squadron's flagship aircraft that had my name rendered on the canopy rail and a hand-painted falcon on the tail. She had taken great care of me until the end. I watched her explode as she hit the ground, thankfully in an unpopulated field far from any nearby villages.

My training immediately kicked in.

There is a Garth Brooks song with the line, "sometimes I thank God for unanswered prayers." As a cadet at the US Air Force Academy, I dreamt of joining the parachute team. Unfortunately, cadets needed at least a 2.6 GPA to qualify to pull their own ripcord. Those of us barely hovering above a 2.0 got to learn how to jump with the US Army at Fort Benning, Georgia, with a static line that pulled the chute out as soon as you left the aircraft.

I was always reminded of a speech President George W. Bush gave at his alma mater, Yale, in 2001. "To those of you who received honors, awards, and distinctions, I say well done. And to the C students, I say you, too, can be president of the United States."

The same can be said about becoming chief of staff of the US Air Force.

The fourth jump at Fort Benning was a full combat gear jump at night, at low altitude. If I had gone through freefall training at the Air Force Academy, I would not have been as prepared as the Army made me for my night, low-altitude, full combat gear jump over enemy territory.

Thank God for unanswered prayers.

I steered to an open field and landed exactly as I had been taught by the "black hat" instructors at Fort Benning. They used to yell through their bullhorns, "Feet and knees together and don't look down!" Once on the ground, we were expected to grab all our gear and run at a sprint to the buses to pass the jump.

It was all so familiar.

I hit the ground, grabbed my gear—just as I did 20 years earlier at Fort Benning—and sprinted for the tree line. When I hit the tree line in the dark, the ground disappeared below me and I fell forward on my inflated raft and rode like Indiana Jones to the bottom of a ravine where I landed in a heap of parachute lines, material, and assorted gear. Ah well, good news was I was still uninjured. My helmet saved me when I came to a stop headfirst.

After burying my gear, I made a call on my survival radio and Jammer Kavlick, the chief of plans whose wing I flew on during this check ride-gone-wrong, answered immediately. I told him I was okay and that I was going to move toward higher ground for the rescue. To save battery life, I also told them I'd be "off the net" for the next hour. In all honesty, I had no idea how long I might be on the ground and needed to preserve every ounce of battery I had.

I started making my way toward some high terrain that I had spotted earlier from the cockpit and was just about to emerge from the ravine when I heard voices. Three armed young men were walking directly toward my position. I hit the ground and slithered my way quietly back down into the ravine. I quietly reached for my 9 mm Beretta pistol in case they discovered my location. I would have the advantage of surprise since they had their weapons slung over their shoulders. When I reached for my weapon, it was missing. It must have flown loose when I ejected from the aircraft. So, I lay in the ravine until I knew they were long gone.

My guardian angels were working overtime.

As I emerged from the ravine, I realized I did not know if I had successfully made it into Bosnia before I punched out. The gauges in the cockpit were unreadable due to smoke and shaking, which meant that I didn't know my position. If I was in Serbia, then the greatest threat was the enemy. If I had made it to Bosnia, then the greatest threat was stepping on a land mine put there by the Serbs. I thought about what a tragedy it would be to survive a shootdown only to get blown up by a land mine.

I carefully maneuvered on the edges of forested areas but only stepped on plowed earth where a farmer would have dug up a land mine if it existed. Turns out I was still well inside Serbia, which meant that being captured by the enemy was the threat. As a squadron commander, I'd be a nice juicy catch for their exploitation. They were out, in force, to get to me before the rescue team.

The race was on.

After an hour evading the enemy, I found what looked like the best spot for a pick up. As I arranged my supplies, I remembered another Desert Storm experience.

One of my squadron buddies had lost his engine over Iraq and needed to eject. Captain Spike Thomas spent six hours lying underneath his raft in a low gully in a desert watching the Iraqis searching for him on the horizon. I asked him how he kept his cool all that time.

He shared that periodically he would reach out from under his raft and scoop up rocks. Using his red flashlight, he picked out the best ones to bring home as souvenirs for his children. I thought then, that is father of the year.

Now, it was my turn.

I had this nice pile of rocks I was collecting for my daughters when I heard footsteps breaking branches in the ravine below my position. I hit the ground again and listened. Slowly, the noise made its way toward me, and I determined that it was likely an animal as opposed to a human.

When it got close enough that I considered it a threat, I jumped up and grabbed a handful of those rocks and threw them at it. Whatever it was reared up and growled at me. Jesse Owens wouldn't have been able to keep up with me as I left that spot to find a better location without company. While I believe to this day it was a mountain lion or bear, my fellow fighter pilots are convinced it was a Serbian field mouse.

Fighter pilot humor.

Due to some complications with my electronic gear, the rescue team that launched out of Tuzla Air Base in Bosnia went to the wrong location. Luckily, Jammer had recorded my location on his radar and was steering them in my direction. I eventually heard chopper noise and vectored them toward me by closing my eyes and listening to their engines. A pair of A-10s had taken charge of the rescue operation and relayed all my calls.

"Come east . . . now south . . . a bit more east . . ." I radioed.

I knew the enemy was likely monitoring the calls, and this kind of specific information would only make their job easier if they were indeed monitoring our frequency. I could make them out on the horizon searching for me. Now was not the time for hesitation. They were close by, and the sun was an hour from rising. It was now or never.

I saw the formation of three rescue helicopters off to my left and directed them to land in the field in front of my tree line. I had placed a strobe light with an infrared cover at a good landing spot in front of my location that could only be seen with night vision goggles. But the helicopters needed to know I was the downed Airman and not the enemy preparing an ambush.

At the Air Force Academy, we were required to rate the performance of the dining hall staff after every meal. The rating form gave a list of options but the right answer to the questions was always: "fast, neat, average, friendly, good, good." Every cadet was taught to answer the form in that way, every day, at every meal. I took my chances that one of the crew was an Air Force Academy grad.

Keying the microphone on my radio, I called in, "Fast, neat, average." One of the pilots came back immediately with, "Friendly, good, good."

I was the guy they were looking for. It wasn't an ambush.

Thanks again to my guardian angels.

As soon as the chopper landed, I came out of the tree line and met the lead pararescue Airman, Staff Sergeant Jeremy Hardy. Sergeant Andy Kubik had taken up a defensive position to the right and Senior Airman Ronnie Ellis to the left. As soon as the helicopter was on the ground, the enemy came racing toward us firing madly hoping for a golden BeeBee to hit. As I met Jeremy Hardy, we locked eyes and he shouted, "Let's go!"

I climbed into the cockpit and was quickly pushed to the deck as the pararescue team climbed on top to protect me from enemy fire. They had body armor; I didn't, and I was their precious cargo. We took off and immediately had to maneuver aggressively to avoid enemy fire. The gunner, Staff Sergeant Rich Kelly, fired his weapon to suppress the enemy continuing to race toward us.

These guys risked everything to save a fellow American. I was headed home to my family for one reason, and that was the courage and dedication of this incredible team. They were not going to quit until I was back in friendly territory.

Guardian angels of a different sort.

All told, we took five rounds in the helicopter that night, including one that passed mere inches from the main hydraulic line that would have forced us to land and reposition into another helicopter in the formation. That would have been tough to do with the enemy closing in. Meanwhile back at home base, my wing commander, Brigadier General Dan "Fig" Leaf, was closely monitoring the rescue effort. As soon as he got confirmation that I was safely out of Serbia and enroute back to Tuzla Air Base in Bosnia, he called Dawn.

"Dawn, I have good news," he repeated about a dozen times. "Yu-Chu (his wife) is on her way to see you and should be there in about 30 minutes. I repeat, she is bringing good news."

Dawn had not slept since she had awoken the moment I was hit, so the call did not come as a total surprise. She believed someone in the squadron had gone down, and as the commander's spouse, she was going to join Yu-Chu to go visit the family to deliver the news. Dawn had done this before with Nancy Diehl when we lost Captain Dale Cormier during Desert Storm. She was praying she would perform as strongly as Nancy. And because Fig told her it was "good news," she was a bit less apprehensive. She fully believed she was going with Yu-Chu to tell another spouse that their husband had been shot down, but it was going to be okay.

When she opened the door, she saw Yu-Chu walking up the path with our squadron chaplain and several others. Seeing the looks on their faces, she knew it wasn't another pilot. She knew then that it was me.

"Dave was shot down and rescued and is on his way home. We don't know his condition, but we'll get a call when it is time to go to the base and greet him." Dawn already had the coffee pot on and invited everyone in to wait.

Her only experience with someone coming home from a shootdown experience was at the end of Operation Desert Storm. She was watching with other spouses as my fellow F-16 pilot, Captain Bill Andrews, was carried on a stretcher when he came home from Iraq. She was fully expecting to see the same for her husband.

Back at Tuzla Air Base in Bosnia, I had the chance to high five every member of the rescue team before loading on a waiting C-130 cargo airplane for the flight to Aviano. I had some small wounds from shrapnel in my right hand that a flight doctor stitched up during the flight. He wanted to wrap me elbow-to-fingertip, but I managed to talk him into a few Band-Aids.

When they got the call to head to the flight line, Dawn woke up our daughter, Dani, and told her that Daddy's plane had been shot down, that he had been picked up and was being flown home. Dawn still didn't know whether I had been badly hurt but was aware that even successful ejections could cause terrible injuries. In the car with our chaplain, Father Marty Fitzgerald, he, Dawn, and Dani prayed together.

When I walked down the steps of the C-130 at Aviano Air Base, a cheer went up from the hundred or so gathered to welcome me home. Fig was first to give me a bear hug and remind me, "You know, I never got shot down when I was the commander of the Triple Nickel."

More fighter pilot humor.

Then there was Dawn—my high school sweetheart—with Dani. Diana was still at her sleepover and would be our first stop on the way home. What an incredible moment it was to be surrounded by the love of family and friends.

After hugs from Dawn and Dani came a memorable moment with Jammer and the rest of the squadron pilots in our formation. In many ways, their experience was more harrowing than mine. Their commander was alone on the ground surrounded by the enemy, and, until the very end, they didn't know my situation because we were avoiding radio chatter.

It came as a giant relief when they saw me walk off the C-130 versus being carried out on a stretcher.

Before leaving the flight line, I pulled Fig and my immediate boss, Colonel Eb Eberhart, aside.

"I know you probably haven't thought about this, but I have. I'm not a young captain that just got shot down. I'm the squadron commander. And I need to get back into the air quickly to put this behind us. I'd like to go home and get some sleep and fly tonight."

Fig looked at me and then at Dawn.

"If she is okay with it, I am, too."

Dawn was on board and knew how important it was to get back on the horse. But our daughters were not so inclined. Our first stop after leaving the flight line was to pick up our youngest daughter, Diana, so I could be with both of my girls. They didn't completely understand what had just happened, and they needed to know that everything was going to be all right.

I had a dilemma. On the one hand, I owed it to my squadron to get back into the fight. On the other hand,

I owed it to my daughters to make sure they understood what happened and that Daddy was okay.

This was not in the squadron commander handbook.

I took that night and the next day to reassure them before I flew the following night. No big fanfare. I just wanted to quietly get back in the air with a message to all that it was no big deal. I had met several of my dad's friends from Vietnam who had been shot down more than once. After their rescues, they quietly got back in the cockpit and returned to the fight. No book tours, no press engagements, just back to work. I had an opportunity to honor them by acting as they did.

We briefed the mission. But before I stepped out to my newly designated flagship, I pulled Jammer Kavlick into a briefing room.

"We never debriefed your check ride," I told him. "So, you flew the formation to a place where your wingman got shot down," I said. "That's not a great start."

He looked down and nodded slowly.

"Then you single-handedly led an all-night rescue effort that brought him home to his family," I continued. "Not a bad recovery."

He looked up at me and I handed him a completed Form 8 recognizing him for the highest score one can earn on a check ride: "Exceptionally Qualified."

As I went through the startup sequence for the next flight, I waited inside the hangar to give the call to taxi to the runway. When I rolled out into the night, I saw a sight I will never forget.

Lined up along every taxiway were folks from across the base. They had all come out to wish me well. Active duty, Air National Guard, Air Force Reserve, spouses, family members, civilians, and both American and Italian workers from across the base. Word had gotten out informally. Each of them gave a thumbs up, a wave, or a head nod as I taxied slowly by.

In the Air Force, we're all family.

The rescue team and I have a lot of fun at reunions. Especially when the moment comes to deliver my annual bottle of single malt scotch to mark the occasion. "I thought you guys were protecting me in the helo until I realized we were getting shot at from below," I joke. "Maybe I was body armor for you."

It is not lost on me that I am writing this story today as a husband, father, and now proud grandfather because of the courage, discipline, and persistence of these incredible men who risked everything they hold dear to bring me home. I will forever be indebted to them for their bravery.

Sometimes, unforeseen setbacks in life store the seeds of a future success that we just can't fully see in the moment. As a leader, it is inevitable that we will get knocked down. But that's precisely when it is most important to get back up. It is when we must use the moment to teach others how to act when they get knocked down. The true test of a leader is not how high we jump, it is how high we bounce.

Get back up.

Chapter 3

Collect Tools

Heather

The Snap-On Tools salesman's truck pulled into our dirt driveway and stopped near the path that led to our treehouse. The bright red panels with their piano hinges looked just like the toolbox in the garage where Dad spent so much of his time.

When I was a child, we lived in the last house the snowplow reached during the long winters on the outskirts of a rural New Hampshire town near the Vermont border. My Dad was a pilot and a mechanic, like his father before him. In the summer, when he wasn't flying the Boston–Miami route for Northeast Airlines, he was likely working on his Volkswagen Beetle or my mother's Jeep in the garage or building his experimental open cockpit biplane in the den.

Whenever a vehicle came down our road and stopped at our house, it was an event. On any summer day, a few people might pass by on their way to the state park at the end of the road. During hunting season, men with orange caps and dogs sometimes crept along, looking for land that wasn't posted.

While the arrival of the rural route mailman in his beat-up station wagon each morning was a big deal for us three kids, the arrival of the Snap-On truck every month or so was like Christmas for Dad. We knew better than to interrupt when the Snap-On Tools guy was in the driveway.

We'd sit on the back porch steps under the window of the bedroom we shared and watch, usually shoeless and wearing hand-me-down T-shirts and shorts with our legs covered in mosquito bites from adventures in the woods that surrounded us.

Dad would come out of the garage with a smile on his face, clad in a plaid cotton shirt, wiping his hands on the handkerchief that my mother taught me to iron for him.

Approaching the truck, he and the traveling salesmen would chat for a while and laugh. Eventually, Dad would reach in his back pocket and pull out a worn leather billfold, flick it open with his thumb, search inside, and hand the man a bill. Smiling, the salesman would swing open the big red door on the back of his truck and hand a tool to Dad.

As the truck backed out of the driveway, we'd scramble to the garage and follow Dad back to his work bench.

Dad would hold the new tool in his grease-stained hand and admire it before putting it in his box and, with a glance at us three troublemakers, return to whatever project he was working on that day.

Eventually, it occurred to me.

Dad didn't need that tool that day. He didn't use that tool right away.

My Dad was a craftsman at his work. And he gathered tools for his box long before he had use for them. He probably didn't even know when or exactly how he would use that tool, but that wasn't the point. He collected them so he'd have them when he needed them.

Craftsmen collect tools before they need them.

Many times, as a leader, I have found myself in novel situations I never anticipated.

I never planned to become the cabinet secretary for the Children, Youth and Families Department in New Mexico. Yet, I became responsible for a gang-infested juvenile prison system that was dangerously overcrowded and a foster care program that was under a federal consent decree for failing to get forever homes for children.

Rather than be paralyzed by what I didn't know—which was a lot—I looked in my toolbox.

As a small business owner, I had learned about the Malcolm Baldrige approach to quality management and been intrigued by it. While its principles like being data-driven, customer-focused, and process-oriented worked for my small business, these also seemed applicable to the larger institution I found myself leading.

But the most important tools in my box I had collected as a junior member on the National Security Council staff where I had watched President George H. W. Bush frame issues for public discussion, communicate quietly with key allies, and advocate effectively with a legislature controlled by the other party.

It was January of 1990. In the wake of the Iran-Contra scandal, the President's National Security Council staff was quite small. At the age of twenty-nine, I was the youngest member of the staff, and conventional forces in Europe was the bulk of my portfolio. The Berlin Wall had just fallen two months before, something I didn't think I would ever see in my lifetime. The President was determined to convince our allies to allow Germany to reunify in the center of a European continent that would be "whole and free."

January's big event in Washington is the State of the Union speech. That year, President Bush used his address to Congress to call for deeper reductions in conventional forces in Europe through arms control negotiations that were already underway.

We had been working quietly on the initiative and the President's speech for several weeks. Deputy Secretary of State Lawrence Eagleburger and Deputy National Security Advisor Bob Gates had taken a secret trip to Britain, France, Germany, and Italy in the days leading up to the speech to confer with allies. And, when the news started to leak the day before the State of the Union, Secretary of State James Baker and Secretary of Defense Dick Cheney worked the phones with allies and key Congressional leaders to give them the details. Just before his speech, President Bush called Soviet President Mikhail Gorbachev to talk through the idea with him.

I was the staff member behind the scenes who knew the substance of the negotiations. I was writing talking points, preparing briefing books, taking notes on telephone calls, and writing memos to capture the gist of those calls. It gave me a front row seat on how senior leaders communicated and built consensus in advance of an initiative.

I also drafted questions and answers to prepare Marlin Fitzwater for what the media might ask in the late afternoon pre-brief before the President left for his command performance at the Capitol. Marlin looked at the draft and asked, "Who wrote these?"

Someone in the room ratted me out.

"They're really good," he said. I wasn't in the room, but, if I had been, I would have exhaled after holding my breath. "Get her down here for the background briefing for the press."

My office was in the Old Executive Office Building across the street from the White House, and I rarely came to the West Wing during business hours. I had only seen the White House Press Room when it was empty, while giving after-hours tours to visitors. Now, on short notice, I was the nameless "administration official" standing behind the podium answering questions from the White House Press Corps on the President's initiative to reduce conventional forces in Europe.

After I survived my first and only time in front of the White House press corps, and exhausted after several weeks of intense work, I took the Metro home to my apartment on Connecticut Avenue and listened to the President's speech to Congress on my radio in the darkness of my living room.

President Bush was famous for keeping his initiatives confidential. The joke on the NSC staff was that you could survive the mistake of compromising state secrets, but don't ever ruin the President's surprise. That was a firing offense.

Until I heard it on the radio, I didn't know that the initiative on conventional force reductions in Europe was the centerpiece of the State of the Union. It would be the headline in the papers the next day. I hadn't seen the whole speech—only the little bit of it that I needed to work on.

Sitting in the dark, listening to the sustained applause from the Congress, I couldn't help but be a little proud. I enjoyed being the low-key staff member who got stuff done. I knew that the bipartisan support and the warm reception from allies were not an accident. The quiet communication in advance had built support, and now the President was mobilizing the support of the American people through his most important annual message to the Congress.

But it wasn't just public communications that I learned from my time in Washington.

In subsequent months, as I worked with people in agencies across Washington to drive forward the President's initiative to reduce conventional forces in Europe, I was particularly impressed by a mechanism at the State Department. Secretary of State Baker had a small team he called his policy planning staff that brought focus to high-priority initiatives the bureaucracy might otherwise smother. It worked well for him and seemed to be a good model to overcome bureaucratic resistance.

Being on the National Security Council staff gave me a front row seat to see the bureaucratic inertia for which Washington is famous.

Every Friday afternoon at 4 p.m. our group of less than ten staff members that worked on defense policy and arms control would gather in our boss's office in the Old Executive Office Building for our weekly "whine and cheese" session. We would rotate who bought a couple cheap bottles of wine, and, as we finished up the week's work, we would laugh about the most outlandish bureaucratic problems. Our boss, Arnie Kanter, who would later become Deputy Secretary of State, had a great sense of humor. One of his favorite lines to make fun of civil servants who might suggest that some policy directive from the President wasn't exactly to their liking was to say, "Well, yes, he is the President. But you're right. It's just one man's view."

We would all laugh at the arrogance of mid-level civil servants who seemed to see the President of the United States as just another bureaucrat.

The Washington bureaucracy could slow anything down. Having arrived at the National Security Council staff from an assignment at the US Mission to NATO in Belgium, I quickly learned that getting Greece and Turkey to agree on something was easier than getting people in the bowels of the Defense Department and State Department on the same page.

And a few years later, I had become the cabinet secretary responsible for the welfare of children in a sprawling state in the great American west.

Creating cells of high performers to drive change and organizational effectiveness, communicating respectfully and in advance with stakeholders, prioritizing and focusing on key initiatives—these were tools gathered long before in a very different situation, but I thought I could use these tools to help improve the state's seriously broken juvenile justice and child welfare systems in New Mexico.

Once in the seat at the Children, Youth and Families Department in Santa Fe, I streamlined the organization and eliminated a whole layer of administrators. Then, I created a policy planning staff and brought into it a mixture of analysts, policy experts, and public servants. We started doing a deep dive on juvenile justice, including a massive data project that looked at the records of every juvenile in custody on a typical day. Heeding the Baldrige framework, we would be data driven. We wanted to know not just how many kids were in custody but what their characteristics were so that we could expand the system in a way that would be both cost-effective and meet needs.

Drawing further from the Baldrige approach to quality, we focused on understanding and improving the whole system rather than just bandaging pieces of it.

The juveniles who were locked up were going to be back on the streets in their communities again. We looked nationwide for programs that had records of reducing recidivism. There was one called the "Last Chance Ranch" in Florida, which took nonviolent repeat offenders into a work camp with positive adult role models. When our study found that we were incarcerating nonviolent offenders in overcrowded facilities with violent teens, we knew that this model might be part of our answer.

We did a second study on juveniles to look at the social, educational, and psychological characteristics of repeat offenders and those who did not offend again to develop a simple, validated risk tool for assessing the likelihood that a teen would reoffend.

It's no surprise to most people that the severity of the crime committed on its own isn't a very good predictor of whether a juvenile would reoffend. If a young teen was behind in school, involved in a gang, had a parent in prison, and shoplifted a candy bar, then there was a high risk for reoffending. That teen needed more than "warn and release" because it was just a candy bar. So, we identified and strengthened programs in communities that gave juveniles at high risk of reoffending more attention, even if it was mandatory after-school study hall, community service, and checking in periodically with a juvenile parole officer.

We developed a package of initiatives to reduce overcrowding in juvenile prisons by adding specialized programs that had worked elsewhere.

I'll always remember the advice of the chief psychiatrist of the Department. "Almost all adult criminals were juvenile delinquents," he said, "but only about half of juvenile delinquents become adult criminals. The problem is, you just don't know which half."

Ultimately, we wanted to increase the probability that we could get young people to turn their lives around before they threw them away and sent all of us the bill.

We called our plan "Restoring Justice," and, drawing on tools I learned from President Bush about public communication and engaging a not-always-friendly legislature, I went on the road to all thirty-three New Mexico counties explaining it to law enforcement, legislators, newspaper editors, and judges. The governor ultimately included it in his budget, and, after lots of hearings and some negotiation, the legislature funded it and we implemented the changes.

But the work didn't stop there.

We then turned to the problem of foster care and finding forever homes for more kids. Again, we drew on the Baldrige tools and took a systems approach, analyzing everything from the database systems to track reports of abuse to partnering with churches to recruit more foster and adoptive parents. And every child in care would have a permanency plan with deadlines for decisions.

Like we had with "Restoring Justice," we created an initiative that explained what we were doing and why. It was called, "One Kid at a Time," and we went on the road across New Mexico engaging legislators, child welfare advocates, children's court judges, and foster parents.

In this case, too, after some negotiation, the legislature approved our initiative.

Gradually, the systems improved. Overcrowding in juvenile facilities declined, and effective programs increased for juvenile delinquents. We opened a work camp modeled on the "Last Chance Ranch" and separated violent kids from nonviolent ones. In foster care, more kids got forever homes with deadlines for permanency plans rather than just growing up in foster care.

My last act before leaving my position as Secretary of the Children, Youth and Families Department was to sign the end of the federal consent decree that had controlled the department and its foster care operations for the previous eighteen years.

The tools we used to get us to that point I had gathered years before. And, when I gathered them, I had no idea when or how I would need them.

Neither will you.

Neither did my dad.

Have a toolbox. And throughout life, collect tools through experiences and deliberate study.

A day will come when you'll need something. You can go to your red box in the corner of the garage next to the work bench, look inside, poke around a bit, and find something that just might work.

Collect tools.

Chapter 4

Find Your Purpose

Dave

Despite all the great schooling I received from North-western Preparatory School, my first two years at the Air Force Academy were not going well. I struggled to maintain a C average and spent months under restriction for violating various rules and regulations. On the athletic field, I was slow.

I was struggling with my purpose.

Why had I gone to the US Air Force Academy when my true passions were music and the outdoors? Why was I committing myself to defending a country I didn't know that well, since most of my childhood was spent in Europe? Whose dream was I really pursuing: mine or my dad's?

I decided to write a paper for an English class my fourth semester at the Academy entitled, "What Is My Life's Purpose?" It described the questions I was pondering and what I would do if I could just get a year off to figure it all out. It somehow got to the dean of students about the same time he received a presentation from a visiting professor describing a sabbatical program they had started the previous year at the Naval Academy.

I was called into the dean's office and asked what I would do if given a year away from the Academy. I told him I'd start with joining a summer program at Philmont Scout Ranch in New Mexico as a back-country ranger followed by joining the band of my musical hero at the time, Harry Chapin. (As it happened, a cousin of my academic adviser was in the band and offered to help me get a position as a roadie.)

Harry had entered the Air Force Academy in 1962 and left after basic training to pursue his dream of a career in music. As a tribute to his short time there, he came back every year to give a free concert to the cadet wing. After listening to him during my first two years, I became a lifelong fan.

It was the perfect scenario. I would pursue my dream of becoming a musician for a year, all while determining what path my life should take.

In other words, I would find my purpose. The dean offered to let me join nine other cadets facing similar challenges as a test group to see if a sabbatical program was a good fit for the Air Force Academy. I would resign my commission, but the form would not be processed until I called back after nine months away. If I wanted to return, there would be a spot waiting for me in the class of 1983.

So, I left.

The summer at Philmont was magical. A ranger's life is all about leading groups of young scouts and their adult advisers into the back country in the mountains of northern New Mexico for the first three days of their trek. I'd teach wilderness survival skills, then leave them on the fourth day to hike back on my own. I'd often stay overnight at various camps where evening campfires were filled with songs and stories.

I never traveled without a guitar strapped to my backpack, and there was nothing I loved more than joining other rangers to lead singalongs around a blazing fire under the stars. I even wrote a few songs that are still sung to this day around campfires at Philmont. It was heaven on earth.

I was well on my way to finding answers to the questions that had brought me there.

Except I hadn't heard anything back from Harry or the band. This was long before the internet and cellular phones, and my adviser had moved on. My lead on being a roadie for Harry Chapin was nowhere to be found.

Now what?

How would I determine if music and the outdoors were my true destiny? Then a fellow ranger shared a book with me he had recently read called *A Walk Across America*, by Peter Jenkins.

Jenkins had backpacked across the entire country and wrote a book about his experiences. It captured my imagination.

This is it, I thought.

I had already mastered the art of hiking with a guitar, and the summer at Philmont only reinforced that writing and singing songs with others were passions. But I only had six months left to decide whether I would return to the Academy.

I couldn't walk across America. I had to ride.

I jumped on my ten-speed bicycle, packed with what I needed as a ranger, ready to travel the back roads of America, try out my music in towns along the way, and see where it all led. Since I no longer had any specific destination in mind, I chased the weather.

It was August in Texas, so I needed to head north. Living on a budget of $5 per day, I got very good at making oatmeal in the morning, a sandwich for lunch, and

Top Ramen noodles for dinner. Not the best diet but a fine one when you're 21 years old and in prime shape.

One of my first major stops was in College Station, Texas—home of the Aggies. I pitched my tent under a large live oak tree on campus and showered in the gymnasium. In the afternoons, I'd grab the guitar and lead singalongs near the student union. Students and professors tossed coins and bills into a dish after joining me on the grass.

$150 later, it paid for another month on the road.

Seeing members of the corps of cadets in uniform mixing freely with civilian students was quite different from my Academy experience where everyone looked the same. Here, there was a quiet mutual respect between uniformed cadets in the corps and civilian students. They shared a common value system that tied all Aggies together.

This was the immediate post-Vietnam era. Respect for the military from students my age was something new. I was intrigued.

As I headed north on country roads, I met people who wanted to share their stories and hear mine. I found that stopping by the local library and asking to read about the local history often led to village seniors coming by to share stories of their town. Local librarians know just who to call.

At the end of a discussion, many would offer me a place to stay. Strangers until that moment, they welcomed me into their homes with their families.

I was meeting the Americans I swore to defend when I took the oath at the Air Force Academy field house in the summer of 1978.

In the center of Missouri, I passed what looked like roadkill until a baby puppy raised her head and yelped as I pedaled past. She was literally petrified as trucks and farm vehicles passed by on both sides.

I got off my bike, ran over, stopped traffic, and scooped her up. She was shivering. I put her delicately inside my shirt. At the next town, I found a vet and showed him my new little friend.

"Folks dump pups all the time," he said. "Just leave her here and I'll take care of it."

"What does that mean—*take care of it*?" I asked.

"I put these dogs down every week unless someone is willing to take them. Leave her with me," he said, "and I'll do my best to find a home. If not, I'll take care of it."

I looked down at her and, I swear, she cocked her head as if to say, "Well?"

"If I decide to keep her and find her a home," I asked the doc, "would you give her the shots she needs? I can only pay you a little."

He looked at me quietly.

"I will," he answered. "On the house."

I named her Minibear after the chipmunks at Philmont who could do as much damage to your backpack as a bear if you let them in. She started traveling in my shirt but soon learned how to sit at the back of the bike on top of my tent and sleeping bag.

We were quite a pair.

I planned to find her a good home and family along the way. We became inseparable. And my diet suited her just fine.

There was no reason to use a leash given our lifestyle. When Minibear wasn't with me, she would roam about on her own but always come back to our bike. During a stop at Southern Illinois University, I was around the corner from where I left my bike talking with some students when I heard her bark for the first time.

That first little bark was intense.

I rounded the corner in time to see her chasing a young man away from my bike. Others told me later he was trying to steal some of my kit when Minibear went after him.

My little savior.

Rather than find her a new home, I decided her home would be the same as mine for the rest of my journey that summer.

Minibear was a magnet and conversation starter in every town I went through. People stopped me to ask about this unusual little dog who was so comfortable riding on the back of a bike.

She opened doors to meeting the kindest people in every state. Ordinary Americans from all walks of life took me, a stranger with his traveling companion, into their homes. If I was a drifter, at least I was a drifter with a puppy.

In Marion, Kentucky, I stayed in the home of two elderly ladies who ran the local jail. They were elected officials who took care of their occupants with home-cooked meals as they awaited trial. Minibear loved running into each of the cells to play with the prisoners.

In Bowling Green, Kentucky, I borrowed a guitar from a local pawn shop because I had sent mine back to my then-girlfriend (now-wife), Dawn, on a Greyhound bus. I needed to lighten the load, which now included a growing pup in addition to my gear. Minibear and I got a gig at a local bar. I strummed and sang for dinner and tips.

Three nights later, we had another $150 to our name. But really, she was the star of the show.

Off we rode.

I awoke one day in the hills of Tennessee to the sound of tires outside my tent. When I emerged, a farmer was standing behind his truck door with a double-barreled shotgun aimed right at me.

"Who the hell are you and what you doin' on my land?" he demanded.

In perfect timing, Minibear came out of the tent and wagged her tail as she ran up to him. He looked down at her and chuckled.

A drifter with a puppy.

"It ain't safe to camp out on someone's land here without askin'," he drawled.

"I'm sorry sir. I'll get packed up and will be off your land right away," I answered.

At this he just stared at me as he reached down to pet Minibear.

"You eaten breakfast yet?"

"No, sir."

He took me and Minibear home in his truck where we enjoyed the best breakfast of fresh eggs and bacon I remember. While we ate, his two children played with Minibear. We could hear their laughter as she ran around their furniture. I stayed with them for several days and worked on their farm.

As the days went by, I came to know the America I would defend if I returned to the Academy. I was finding my purpose.

Still following the weather, Minibear and I continued over the Smoky Mountains then down the East Coast to

Florida. Two of my high school buddies were in college in Tallahassee, so I pedaled into town to stay a few days with them.

We were outside the house sitting around a fire pit with the guitar when I heard a car screech and a loud yelp (like her first when I found her on the road in Missouri). Minibear had been wandering around like she usually did; however, this was our first time in a big city. I ran to the car, the driver got out, and there was my little buddy under the front tire.

Alive. Barely.

We raced to the vet, who quickly put an IV in her and did an examination. Things did not look good, but she was a tough little pup. I stayed with her as long as I could before the vet asked me to leave the room. I took off my shirt and wrapped her in it so she would at least have my scent around her as she fought for her life.

Early the next morning, I raced back to the building. The vet met me at the door.

"I'm sorry, Dave. Minibear didn't make it."

My travel buddy was gone. We had been together for six months. What a blessing she was. How I would miss her for the rest of my journey. Even now, to this day.

It was somewhere on the coast in Alabama where I had to make my decision to either return to the Academy or have them process the resignation letter I signed before leaving.

I'd left intending to make my way in the music business while pursuing my passion for the outdoors. Instead, I got to know the country I had taken an oath to defend.

Pondering the decision, I concluded that the people I had met on this six-month journey were the best of

America. These were strangers who told me their stories and took me into their homes. People whose families were woven together in communities, all of them caring for each other and each of them pursuing their own dreams. These people were worth defending.

It was no longer me pursuing Dad's dream. It was me pursuing mine.

I had found my purpose.

At the next library, I stopped to ask if I could borrow a phone to call the Academy. "I'm coming back," I told them.

We all have a unique purpose on this earth. Sometimes, though, the path that leads us there is not the one we planned. As his own story goes, Harry left the Air Force Academy after basic training and made his way to New York where he formed a band with his brother and drove a taxi at night.

He captured this journey in the chorus of his hit song, "Taxi":

She was going to be an actress

I was going to learn to fly.

She took off to find the footlights

I took off to find the sky.

He had found his purpose.

The young woman in the song, Sue, tried her hand at acting but ended up raising a family and discovering true happiness as a wife and mother.

She had found her purpose.

As leaders, finding your true purpose can become a powerful foundation for living a life of meaning.

As for my story, who knew that I'd end up emulating Harry after all. He left military service to find his way into music. I left music to find my way back to serving my country in uniform.

Find your purpose.

Chapter 5

Be Aware of Blind Spots

Dave

The US military was integrated far earlier than most other institutions in our country. In 1948, after Black Americans proved themselves in combat units like the Tuskegee Airmen in World War II, President Truman desegregated the military by executive order. Through Korea, Vietnam, Desert Storm, and the war on terrorism, Americans of every race and creed have served together.

We are a military organization sworn to uphold core values of integrity, service, and excellence. Our organization is also a reflection of the society we serve. Bias or unfair treatment in the wider world would not stop at the gates of our bases, regardless of whether we built a culture that sought to treat everyone wearing the uniform based on merit.

As the Chief of Staff of the Air Force, I knew it was important to ensure I was always appropriately sensitive to any discrimination or unfair bias that would impact our readiness to accomplish our missions. There was only one CSAF, and my job was to ensure all Airmen (active, guard, Reserve, and civilian) truly believed I was *their* chief who represented everyone who took an oath to serve and their families.

As a white male, I also knew the importance of listening and appreciating my own limitations when it came to understanding what others might be going through. Servant leadership is not about being the loudest voice or the smartest person in the room. It's about creating a trusted space for others to have conversations about hard topics and be shown respect. It's also about drawing from examples set by others and lessons learned in crisis.

As a young squadron commander at Aviano Air Base, I was fortunate to work with an extraordinary senior enlisted adviser. Chief Master Sergeant Jimmy Kelly had a way of teaching things that stuck with you. He did so not through lectures or directives but through quiet, powerful moments of insight. One of those lessons stayed with me for the rest of my career and helped me lead more effectively.

It was midday, before one of our regular walks on the flight line to connect with Airmen, when Chief Kelly walked into my office. He held a small box in his hand and tossed it onto my desk. It was a box of Johnson & Johnson flesh-colored Band-Aids.

"Sir, with all due respect, this ought to make you mad because it makes a lot of your Airmen mad," he said calmly.

I picked up the box and turned it over in my hands. It wasn't clear to me what he was trying to point out.

"Sorry, Chief. I'm not getting it," I said.

He opened the box, pulled out a Band-Aid then placed it on his arm. Its soft pink hue stood out in stark contrast against his dark-toned skin.

"Sir, this ought to make you mad," he said again. "It sure makes a lot of our Airmen in the squadron mad." He gave me a knowing look, and we left together for our flight line walkabout. This was his way. Quiet. Thoughtful. Deep. It was a powerful lesson.

I started to get it. As his point sunk in, I also realized that I would never fully understand his perspective as a person of color or be able to fully grasp what he felt along with many others I was serving as leader. For my entire life and career, systems and norms around me were designed by and for people like me. What I had seen as something "neutral" or "standard"—a "flesh-colored" Band-Aid—was, in fact, alienating for many of my Airmen.

I had a blind spot. We all have them, but it's especially critical for leaders to keep this—and their own—in mind.

Blind spots aren't inherently problematic or a failure, but they can become a liability if we don't humbly acknowledge them. If we're not careful, they can lead to repeated mistakes, hinder growth, and damage relationships by perpetuating misunderstandings or missed opportunities for learning and connection.

The lesson I took from Chief Kelly that day was that I could not rely solely on my own perspectives if I wanted to lead effectively. To see beyond my own limitations formed by my life experiences, I needed to surround myself with people who had very different experiences, backgrounds, and pathways than my own. In other words, I needed a team of advisers who didn't look, sound, or think like me and who were willing to offer checks on my perspective.

As we addressed these issues as CSAF, my primary teammate was Chief Master Sergeant of the Air Force Kaleth Wright. He became a powerfully authentic voice for all Airmen, not just Airmen of color, who saw in him a leader who looked like them and shared many of their life experiences both in uniform and out.

His willingness to share his personal experiences created an opportunity for others to do the same. He highlighted for us all that vulnerability and self-awareness are not weaknesses. They are strengths. When leaders appear to take risks and hold a mirror up to problems, they create an environment for others to follow suit. Doing so builds trust, fosters connections, and lays the groundwork for meaningful change.

Working together, we created a space for conversation and invited others into it. They came, they shared, we heard. In setting after setting, we heard stories of perceived bias in promotions, of subtle but persistent discrimination across ranks, and of the pain of feeling invisible in an institution they loved. Young women and officers of different social, religious, and economic backgrounds spoke about being repeatedly mistaken by seniors as enlisted Airmen—a reflection of biases about what an officer is "supposed" to look like. These stories and countless others underscored the work we still needed to do as an institution.

Servant leadership demands more than glorifying the problem. It demands action.

Mostly though, we knew that a rollout of big Air Force–wide programs would inevitably fall short and that problems were ultimately best confronted by local leaders at squadron, group, and wing levels.

When I was a squadron commander, I had the closest relationships with my Airmen and many of their families and could therefore facilitate more candid discussions. As I rose to command larger organizations, however, it became harder to form these very personal relationships just because of sheer size.

This is why Chief Wright and I decided to lead a Facebook fireside chat at the earliest opportunity, followed by a two-hour Facebook Live discussion with Airmen to kick off what we called a "campaign of learning" across the Air Force. We emphasized to commanders that these open discussions were not going to solve the issues in a single forum but rather were best viewed to uncover the blind spots resident in every organization. In doing so, we believed we would create a better understanding of what service in the Air Force felt like. Perhaps most importantly, we ensured leaders at every echelon of command knew we trusted them to rise to the occasion.

Here's what we learned: listening is not passive. It's active and can be transformative. It requires leaders to set aside their assumptions and make a genuine effort to tune in to the pain, frustration, and hope of those they serve. For me, these conversations were a wake-up call reminding me that leadership is not about protecting an image, an institution, or avoiding uncomfortable truths: Leadership is about confronting tough topics with courage, treating others with respect, and making the unit better by understanding and addressing its needs. Leadership is not about having all the answers. Often, leadership is about asking better questions, listening with humility, and acting with determination.

Being a leader who truly represents all entrusted to our care is both a daily challenge and an opportunity. America's parents entrust their greatest treasure to us when they send their children to serve. We owe it to them and to America to continually improve the culture and essence of what it means to live and work in the organizations we are privileged to lead. Admitting to and then proactively addressing blind spots as a leader is a sacred responsibility.

Be aware of blind spots.

Chapter 6

Take the Shot

Heather

It was the summer after my freshman year at the Air Force Academy, and I decided to give up my three weeks of summer vacation to take an academic class. I knew the coming fall semester would be too much for me to handle, especially because I was on the debate team and my partner and I would be traveling to a lot of competitions around the western United States. I needed to get at least one course out of the way or risk drowning academically.

So, I signed up for the required core course in computer science. It would be an entire semester jammed into three weeks. Then, if I survived computer science, I would learn to fly gliders. I was really looking forward to that.

But first, computer science.

I didn't know anything about computers. This was before IBM came out with the first desktop. We were coding on punch cards after writing drafts of code on sheets of notebook paper. They didn't trust introductory students like us to write code on the green television screens in the room next to the air-conditioned expanse that housed the mainframe computer.

There were about thirty of us taking computer science. Of these, several cadets were in summer school because they hadn't passed the course the previous year. Like the rest of the core curriculum, you had to pass this class to graduate.

While I had never been exposed to computing before, I quickly discovered that I was good at it. Computing language made a lot more sense than Russian, which had tortured me for two semesters.

It wasn't long until I was writing answers in code before the instructor finished explaining a question. And, in the few hours a day we weren't in class, I became the tutor for struggling students.

Our instructor, a young captain, knew I was taking the class to get ahead for the fall semester. He suggested that I major in computer science.

I liked computing, but not that much. Besides, was there really a future in this brand-new computer science thing?

In hindsight, I probably should have thought about that decision a little bit longer.

The young captain also suggested something that would change the course of my life.

"If you weren't so busy doing all this leadership and debate competition stuff, you could focus on your academics and apply to be a Rhodes Scholar."

I had no idea what he was talking about. As far as I knew at the time, a Rhodes Scholarship was something sponsored by the Department of Transportation.

But he said I *couldn't* do it if I continued to spend so much time on things other than academics. Perhaps he knew a bit more about me than I knew about myself at the time: if anyone told me I couldn't do something, it was a challenge. I was at least going to find out what he *thought* I was incapable of doing.

At his suggestion, I went up to the top floor of the academic building and found a small one-person office where a former secretary to a department chair, Mrs. Fern Kinion, was set up to talk to students about applying to graduate school. Mrs. Kinion welcomed me. She showed me some three-ring binders with information about graduate scholarships and told me I could come back anytime to talk. I soon learned a Rhodes Scholarship is one of the oldest scholarships in the world granted each year to thirty-two students across America for post-graduate study at Oxford University in England.

I started thinking about graduate school. In rural New Hampshire, the only people I knew who went to college for more than four years were medical doctors and lawyers. The truth was, most Rhodes Scholars had higher grades than I did, although I didn't really know that at the time. I talked to a few of my professors and my debate coach. All of them encouraged me to go for it.

So, I sat down to write an essay about what mattered to me and why. I had never traveled outside of America, and going to graduate school in another country seemed like such an adventure. I was curious and had been exposed to so many interesting subjects in the intense core curriculum at the Academy. But I wanted to go deeper. I wanted to learn more.

Eventually, I told the Academy that I wanted to apply for the Rhodes as well as the Marshall Scholarship, which also funds postgraduate study in the United Kingdom but has more of a focus on excellence in academics and less on the development of leaders who will "fight the world's fight."

The dean didn't think my grades were good enough for a Marshall. He was right. Still, I was one of the leaders of the cadet wing, and the Rhodes valued more than mere bookworms. The Academy wouldn't block me from applying for the Rhodes, but I would have to be screened by a panel of faculty to earn their endorsement. Mrs. Kinion told me that, if I was selected for an interview by the Rhodes committee in my state, I could go home a week early for Christmas break. That seemed like a particularly good reason to apply, so I did.

The meeting with Academy faculty on the selection committee wasn't a pro forma interview in my case. My grades really weren't all that great for Rhodes applicants, which meant the interview was critical.

Five faculty members and the assistant dean greeted me as I walked into a small room wearing my crisp dress uniform and took my seat. After some initial pleasantries, a faculty member asked if I could explain how a wing flies.

I was in luck.

A few weeks before, I had an epiphany in a late-night study session for physics while also taking aerodynamics. I realized that I could derive Bernoulli's equation, which explains force due to the flow of air over a curved surface, from Newton's second law of motion, which says that force is equal to mass times acceleration. These connections between disciplines fascinated me, and, at that moment, the idea was fresh. So, I explained it.

They had other questions, too, but I could tell by the looks on their faces after that answer that they would support my application. Looking back on it now, having served on Rhodes selection committees for many years since, I know their reaction wasn't so much about my ability to do the math but rather my excitement and interest in a question that went beyond what was taught in the classroom. That, and my ability to explain something complicated to people who weren't experts in the field.

At the Rhodes interviews in New Hampshire and then in Boston, I was the applicant that wasn't like the others.

The rest of the finalists had all gone to Ivy League schools in New England, and when I showed up in my cadet uniform with jump wings and a glider pilot chevron on the left shoulder, it was clear that most on the committee had never seen a woman in a military uniform before.

I was different, and I was okay with that. I had always been okay with that.

The year I was selected as a Rhodes Scholar, we all met in New York before flying overnight to England to start graduate school at Oxford.

Of the thirty-two of us in New York that night, thirty-one of us thought we were the one mistake.

The other guy was a jerk.

As a two-time university president, when I talk to my exceptional students today and encourage them to consider applying for scholarships like the Rhodes, Marshall, Fulbright, Mitchell, Schwarzman, or other graduate fellowships, I remind them that my resume doesn't include the long list of things I haven't accomplished in life.

My resume doesn't tell you that Brown University didn't admit me as an undergraduate.

It doesn't tell you that the Air Force Academy didn't support my application for a Marshall Scholarship.

It doesn't tell you about the jobs I applied for that I didn't get.

And it doesn't tell you that I lost a race for the United States Senate, although Wikipedia would.

But what Wayne Gretzky said of hockey is true for me: I've missed 100 percent of the shots I didn't take. It's true for all of us.

So many people are afraid to fail or, more accurately, are afraid to be *seen* to fail. And they don't take the shot.

They don't volunteer for the tough job or apply for something they might not, in their mind, be the best qualified for.

Women, in particular, are notorious for this. Thanks to job sites like INDEED and Zip Recruiter, there is pretty good data that about two-thirds of women won't apply for a job unless they think they meet all the requirements posted in an advertisement. About two-thirds of men will apply if they believe they meet about half of the qualifications. I'm not sure why that is, though I can speculate that a lot of young women are encouraged to be more modest or humble in order to be socially acceptable.

I've hired hundreds of people in my career and can tell you that almost *no one* meets all the qualifications in the job description.

For most jobs, or scholarships, or awards, there is the "Mr. Potato Head" moment, where no candidate has everything, but everyone has something valuable. Most of the time, making the choice involves a compromise. Every time, however, there's a zero percent chance you will get the opportunity if you didn't apply in the first place.

And, sometimes, even if you miss, other opportunities will open because you met someone or learned something interesting in the process that connects you to whatever is next.

That has happened to me countless times.

In early 1994, Albuquerque Public Schools was looking for a new superintendent. They had significant management problems and advertised that they wanted someone who wasn't necessarily a career educator. I had been involved with our local schools as a small business owner and knew the schools needed help. I decided to put my name in and ended up being selected as one of three finalists.

They gave the job to someone else—a more traditional candidate.

During the interview process, I met more people in the community. One was a state legislator and Naval Academy graduate named Kip Nicely. We went out for coffee and got to know each other.

In November that same year, when a new governor was elected, I received a phone call from the incoming governor's chief of staff.

"The governor-elect would like to talk to you about becoming cabinet secretary for the Children, Youth and Families Department."

Kip Nicely had recommended me for the role.

I never would have become a cabinet secretary in New Mexico if I hadn't tried—and failed—to become superintendent of schools.

And if I hadn't been a cabinet secretary, then I probably never would have become a member of Congress.

But long before either of those positions in government, I never would have become a Rhodes Scholar if I didn't take the shot.

Take the shot.

Chapter 7

Own Failure

On Sunday November 5, 2017, shortly after 11 a.m., 26-year-old Devin Patrick Kelley, dressed in black tactical gear and carrying a Ruger AR-556 semiautomatic rifle, walked into the church service at the First Baptist Church in Sutherland Springs, Texas. He opened fire and killed twenty-six parishioners, including an unborn child, and wounded twenty-two others before killing himself. It remains the deadliest mass shooting in an American place of worship.

Dave

It was early Sunday afternoon when the Air Force Inspector General, Lieutenant General Sam Said, called and told me that some of the victims might be associated with nearby Randolph Air Force Base.

But it was worse. The shooter was a former Airman.

From high school through his time in the service, Kelley's mental state degraded. His behavior became increasingly violent; he was incarcerated and ultimately discharged for bad conduct in 2014.

I immediately contacted the commander at Randolph to make sure we were doing everything possible to support the community.

Heather

I was in the loft study of our townhouse in Alexandria, Virginia, early Sunday afternoon when Lieutenant General Said called me with the bad news. He told me the same thing he told Dave: Kelley had been an Airman, discharged for bad conduct. General Said wouldn't be able to confirm until Monday morning, but it looked like Kelley was able to purchase his weapon even though he had been convicted by court-martial for assault against his wife and stepson. That conviction should have been reported to the National Instant Criminal Background Check System (NICS) because federal law prohibits anyone convicted of domestic violence from possessing firearms.

I asked a few more questions, but we wouldn't have the full story until early Monday morning. Dave and I decided to get the team together Monday at 9 a.m. in my conference room at the Pentagon.

Throughout the afternoon, I watched the news coming out of Texas. It was sickening.

Dave and Heather

On Monday morning, everyone who had information or who would have to take action was in the Secretary's conference room: lawyers, public affairs, legislative affairs, the inspector general, human resources, and the Office of Special Investigations (OSI).

Sam Said gave everyone the bad news, that Kelley had been an Airman and our Office of Special Investigations had not reported the domestic violence offense in the NICS. They were supposed to, but they didn't do it. Had they done so, it might have prevented Kelley from buying the gun.

We failed—with tragic and deadly consequences.

We didn't think it could get any worse, but it did.

There had been a Defense Department audit several years before highlighting problems implementing the mandatory reporting law in all the services and recommending improvements. While that report said the Air Force record of reporting convictions to NICS was better than the other services, the recommended corrective actions were never implemented.

We sat there looking at the table, looking at each other, and letting the impact sink in.

Both of us had been in enough lousy situations to know that organizations screw up for all kinds of reasons. Usually, it's not malicious. People aren't trained or offices are short-staffed. The IT systems don't work as they should, or procedures were never written down and the person who knew how to do that job retired. There are lots of reasons. But sometimes, regardless of the reasons, institutional failures have tragic consequences. This was one of those times.

We turned the discussion to what to do next.

We had to review why the domestic violence offense hadn't been reported to NICS and, as necessary, hold people accountable. Lieutenant General Said, an officer we had complete confidence in as a man of integrity, would take the lead on that investigation.

This one case begged a question: How many others had we failed to report? We created a task force to review the records of every Air Force military justice case that should have resulted in an NICS report back to at least 2007 when the law was amended and, if they decided they needed to, back to 1993 when the record check law was originally put in place. We knew this would take months to complete and thousands of man-hours, but it was the right thing to do.

Then we talked about what we should say to the public and the Congress. Those sitting around the table expressed different views. Some pointed out that we weren't in charge when the mistake was made. Our general counsel was traveling that day, but a more junior lawyer was in the room. He advised us to be vague and say we were looking into the matter because of the potential for lawsuits and damages. Others recommended we share the blame across the Department of Defense, since similar failures were uncovered during the investigation into all the services.

We didn't take any of that advice.

As a leader, blaming your predecessors is rarely effective. You are the embodiment of the institution. You wear the coat of leadership, and you can't take it off. You must speak for the institution, and we had to represent our service. It didn't matter that the failures didn't happen when we were in charge; as leaders, we had the watch.

And, while lawyers are valuable to have around, they are there to give advice. They don't decide. We knew that Devin Kelley's conviction should have been reported to the FBI, but it wasn't. Being vague wouldn't change that fact. And throwing our sister services under the bus for being worse than the Air Force wouldn't change anything for the families of Sutherland Springs.

We often came at a problem from different perspectives and occasionally disagreed on an issue; and we both felt this was a great strength in our relationship. We rarely let a day pass without meeting in person or having a call to compare notes, sometimes multiple times a day. On this decision, there was no daylight between us from the very beginning.

We decided to own it.

It was, in the short-term perspective, a hard call. And it was certainly one of the worst days we had during our service together at the Pentagon. Perhaps more than any other event we dealt with as Secretary and Chief, we drew on our common foundation of values developed through early experience at the Air Force Academy. Both of us knew that owning failure was the right thing to do.

We directed public affairs to draft a statement for release no later than 11 a.m.—two hours after our meeting started—acknowledging the failure and committing to full disclosure as we determined how it had happened and what measures we would put in place to fix the system.

It took them a little longer. (It always does.) But, by 1 p.m., we had the statement in the hands of Pentagon reporters.

The Deputy Secretary of Defense also wanted the Defense Department Inspector General to conduct a review. We didn't mind that. He was certainly within his authority to do so, and we knew that the problem extended beyond just the Air Force. But we already knew the important facts. We would own up to our problems and focus on fixing them.

We also expected we'd be well into implementation of corrective action by the time any Department of Defense Inspector General's report was complete. They take forever. In fact, the DOD Inspector General's report wouldn't be completed for over a year—months after the Air Force finished its review and put in place new policies, procedures, and training.

The next day, we sat side-by-side in the Pentagon briefing room and answered questions from the press. The questions were direct and tough, as they should have been.

Kelley's conviction should have been reported, and it wasn't. The shooter was an Airman who had been tried by courts-martial and convicted in an Air Force justice system. It was Air Force OSI that failed to document the case in a way that could have prevented him from purchasing several weapons he had on him the day of the shooting. We told the press we were going to investigate whether there was intentional wrongdoing, and we were going to fix the problem.

You could almost feel a certain emptiness in the room.

There is often the tendency to "deny, delay, and deflect" with the press. Our decision to take responsibility and even to refuse to blame our predecessors seemed to keep the event from turning into the meaningless parry and response that is all too common in Washington, DC.

For the moment, there just wasn't much more to be said.

Heather

In the weeks that followed, we reached out to members of Congress to brief them on what happened and what we were doing to correct the problem.

Senator Ted Cruz of Texas was justifiably direct, asking questions about what went wrong and how, whether people would be held accountable, and what would change. He was thinking about legislation. I advised him that I didn't think more law was needed. The problem was that we hadn't implemented the law already on the books. He couldn't fix that; we had to.

Dave wanted to go to one of the funerals in Sutherland Springs, if the family would have him. He and I talked about that. I agreed completely. While I could have gone too, there was no question in my mind that he was the right one of the two of us to attend. While, by law, the Service Secretary has almost all the authority, the Service

Chief has earned the respect of the American people in a different way. I was a confirmed political appointee. That's important, but it isn't what people think of when they think of the American Air Force.

Dave, in his uniform with four stars on his shoulders, represented something more. He carried the history and values of our service in a visible way. He could best begin to rebuild the trust we had betrayed with the Sutherland Springs community and with the American people. I knew he could represent us better than I could.

Dave

A few on our staff asked if it might be more considerate for the family if I attended the funeral in civilian clothing rather than my uniform. I asked my team to inform the families and the pastor that I wished to attend the funeral and honor the victims in uniform but would certainly change this plan if my attendance would cause further pain to this grieving community. I was told I would be welcome in uniform.

On November 15, ten days after the shooting, I attended the funeral for nine who were killed. While I have lost several Airmen under my command in combat over the years and have sat with many grieving family members, this one was profoundly different.

Of the nine victims we buried that day, seven were family members of a young Airman who we flew home from his deployment overseas to be there. I marveled at his strength as we embraced. The greatest treasure in our nation's arsenal continues to be the young men and women who stand to defend all we hold dear. Like his pastor, this young man was full of grace. What a gift that he had chosen to become one of us.

Also among the victims was the 14-year-old daughter of the church pastor, Frank Pomeroy, who led the service. It remains one of the most courageous and profoundly religious moments during my time as Chief. Not only did Pastor Pomeroy have the strength to stand before the congregation after losing his beloved daughter and parishioners, but he asked all of us to pray for the shooter and his family. Something I will never forget.

Heather

About a month after the horror of that day, the Senate Judiciary Committee held a hearing on background checks. I testified about what happened in Sutherland Springs and what we were doing. I told the Judiciary Committee that the case should have been reported and it wasn't. I took full responsibility and told them what we were doing to fix the problems.

Anyone testifying on Capitol Hill needs to be prepared to be grilled, but I've received far tougher questions on far more benign topics. The senators—particularly those from Texas—had a right to be disappointed and angry with us. We failed to implement a law they had passed to help prevent violent people from buying guns. They knew from my opening statement that we weren't offering excuses, and, truth be told, they couldn't have been more disappointed in us than we were in ourselves.

Our very capable Chief of Legislative Affairs, Major General Steve Basham, or "Bash," as we called him, was a B-2 bomber pilot by profession. He was one of the people in the room on that Monday morning when we all absorbed the terrible news. He'd been with me throughout the weeks of private briefings with members of Congress and helped to prepare me for the hearing. He also sat behind me stoically while the senators asked all their questions. As we walked out of the hearing room, he made a comment that stuck with me.

"I learned something important today," he said.

"What's that?" I asked.

"The outcome of that hearing wasn't determined today. It was determined in the twenty-four hours after the shooting in Sutherland Springs. You and the Chief took responsibility," he said.

He was right. Sometimes, taking ownership may seem like the harder thing to do in the short term. But doing the right thing immediately can turn out to be easier in the long run.

Dave and Heather

As expected, a lawsuit was filed against the Air Force. The Department of Justice handled the case, as they do for all lawsuits against the federal government. While some lawyers might argue that the case would have been easier to defend if we hadn't accepted responsibility immediately, we aren't convinced of that. It might have taken longer, but the outcome would have been the same because the truth would have been the same. And taking ownership made it easier for the families of the victims and the Sutherland Springs community to accept our continued assistance.

After the funeral, Dave met with the mayor of San Antonio to discuss a partnership to ensure we remained committed to this community with every available resource in the months ahead. We received regular updates and kept in contact with the victims' families throughout our tenure together.

When an organization fails, it's best for leaders to take immediate ownership and responsibility. While successes can and should be shared, failure must be owned. In some ways, it defines the adage of "loneliness at the top." But with every failure comes opportunity and the seeds of future success. How the leader acts once a failure is uncovered defines the essence and character of the organization. Taking responsibility and then doing the work to find a lasting solution not only makes the organization better, but it also shows accountability, humility, and commitment that in turn cause the kind of good and decent people you need in your organization to want to follow you.

November 5, 2017, remains one of the worst days in our professional lives.

And yet, we will always be inspired by how the Sutherland Springs community rallied together and responded to this tragedy. It is one of the greatest examples we witnessed of the power of the human spirit.

May God continue to bless the community of Sutherland Springs, a community that represents the best of the Americans we have been so privileged to serve.

Own failure.

Cadet Heather Wilson at the US Air Force Academy (USAFA), 1981.
As a member of the Academy's third class of women, I stepped into a world of possibilities I hadn't imagined as a kid growing up in New Hampshire. USAFA opened doors that I never even knew were there and instilled values that continue to guide my leadership today.

Cadet David Goldfein, USAFA, 1983.
I didn't get into the Academy on my first try and wasn't a standout cadet. Looking back, the experience challenged and shaped me in ways I didn't fully understand at the time. The relationships I built came full circle throughout my career, and the values I learned became the bedrock of my life as a fighter pilot and leader.

▲ **Goldfein with his Peugeot 10-speed bicycle and travel companion, Minibear, 1980.**
Between Academy years, I took time off to bike across the country and reflect. That journey along America's backroads reaffirmed my commitment to return to USAFA, graduate in the class of 1983, and serve the country I had come to know more deeply, mile by mile.

▲ **Goldfein as F-16 Mission Commander during Operation Desert Storm, 1991.**
This marked the beginning of a long series of air campaigns, from the Balkans to the Middle East. We knew the risks and the stakes were high, but that's what we signed up for.

▲

**Goldfein after his rescue, arriving
at Aviano Air Force Base, 1999.**
The entire rescue team from Aviano showed up to
welcome me home. I knew what I was facing that
night—they didn't. In many ways, the uncertainty
made it harder on them than on me.

▲

The rescue team that recovered Goldfein from Serbia where he was shot down over enemy territory, 1999.
I owe my life to this group of heroes—and a bottle of scotch every year on May 2. In this line of work, your life can change without warning. That's why we train, trust each other deeply, and never take a day for granted.

▲

Camp Unity, Pentagon, 2001.

9/11 was our darkest day—and a moment of remarkable unity. In the wake of tragedy, Americans showed up. Camp Unity began in the Pentagon parking lot, built by people who simply wanted to help. Dave was fortunate to have played a small part in bringing together volunteers. Heather, a member of Congress at the time, came to visit and support those aiding in the recovery. We didn't know we were in the same parking lot and would return to the Pentagon sixteen years later as Air Force Secretary and Chief of Staff.

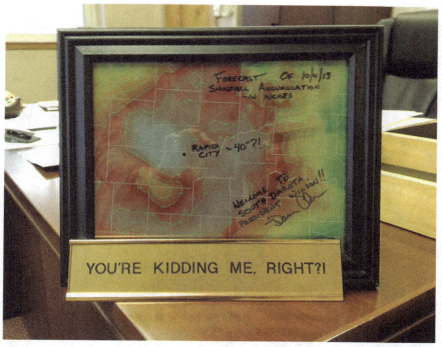

▲

Wilson's desk while president of South Dakota School of Mines in Rapid City, 2013.
A snow forecast prompted my response, "You're kidding me, right?" It was engraved on a nameplate that's followed me to every job since—a fitting reaction for the many moments when someone walks in and says, "Boss, there's something you need to know. . . ." In leadership, surprises are constant, but a bit of humor can encourage good people to bring you bad news and then focus on what needs to be done.

▲

**Tom Friedman of *The New York Times*
and Phil Stewart of Reuters with Goldfein
and Wilson in Afghanistan, 2017.**
It was important to go forward to where Airmen were
deployed and listen to their needs and their stories.
The "battle buddy" on the front of Secretary Wilson's
body armor was made by an Airman who wanted to
talk to her. She still carries it attached to her
briefcase.

▲ **Pentagon Medal of Honor Ceremony, 2018, with Valerie Nessel, John Chapman's widow, and Chief Master Sergeant of the Air Force Kaleth O. Wright.** Honoring John Chapman, our first Airman awarded the Medal of Honor since the Vietnam War, was an unforgettable privilege. It was also a solemn reminder of the courage and sacrifices that define our service.

▶

Wilson, speaking with David Ignatius of *The Washington Post*, 2018. The Secretary of the Air Force is a public figure who interacts with all kinds of stakeholders. A core responsibility of the job was to make our story clear and compelling to audiences beyond the military.

▲

Commemorating the anniversary of the Air Force at the Pentagon, 2017. Marking the Air Force's seventieth birthday was a celebration, but leading its people was the real gift. We strove daily to be worthy of their trust.

▲

Goldfein's office as Chief of Staff of the Air Force.

I often walked visitors through the story of our Air Force using the photos on my wall—none of which were about me. The final one, seen over Mike Holloway's shoulder, was of Medal of Honor recipient John Chapman. Mike later joined the Air Force as a defender.

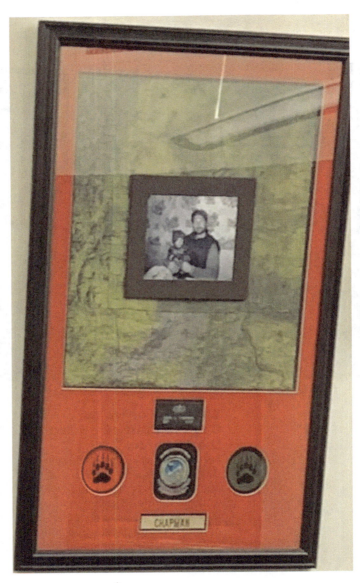

▲
The picture of John Chapman
that hung on the wall behind
Dave's desk during his entire
tenure as Chief of Staff,
2016–2020.

▲
Wilson and Goldfein at a base overseas, 2017.
For both of us, interacting with Airmen was always the best part of the job and what we will remember most.

▲
Wilson flying over Florida with Air Force Special Operations as Secretary of the Air Force, 2018.
I didn't fly in the Air Force, but I'm a private pilot. This Special Operations Cessna had much better avionics than my little Cessna 152. One priority during my tenure as Secretary was to buy things faster and smarter—including avionics upgrades.

▲

Wilson and Goldfein at Wilson's retirement ceremony at Andrews Air Force Base, 2019.
As Secretary and Chief, we often talked about our different roles under the constitution. We worked hard to maintain open communication and a relationship of trust, and we laughed a lot—an underappreciated tool for leaders. As Secretary and Chief, it was important that our subordinates knew that we respected each other and worked well together.

**Wilson presiding at commencement at
The University of Texas at El Paso, 2023.**
Graduation is a day families remember for a lifetime.
It is a rite of passage that reinforces for everyone
present the importance of our mission.

Chapter 8

Trust Your Gut

Dave

It was May 1999. I was out walking the flight line with my chief, the top enlisted leader in the squadron, while watching my pilots and maintainers recover from an afternoon launch. I had flown a mission earlier that day over Kosovo, and now it was time to focus on the support end of the business. There was nobody I enjoyed spending time with more than Chief Jimmy Kelly.

I was his boss; he was my trusted senior adviser. While our partnership never crossed the line to friendship in the civilian sense while we were in uniform, he was the person I was closest to as the squadron commander.

We'd been at war over Kosovo and Serbia for a little over three months, and I was very happy with how my squadron of F-16 fighters—the Triple Nickel—was performing. We were launching and recovering F-16s every day and night and had a battle rhythm I thought was sustainable. The pilots were hitting their assigned targets with incredible accuracy, and the jets were achieving a mission capable rate of over 90 percent. That rate was unprecedented in peacetime operations, and as commander, I couldn't have been prouder.

Since my shootdown and rescue, things had returned to normal. It was just what I had hoped for when I got back in the cockpit and flew the next night and every night thereafter.

As we walked to each hardened aircraft shelter to check in with maintainers, refuelers, production supervisors, and the rest, I could tell the team was focused. Morale was high and people were ready for the long haul if the war dragged on.

When I got back to squadron headquarters, my operations officer was waiting for me. He was obviously concerned about something and asked to talk privately, so we went into my office and grabbed some coffee.

"What's up?" I asked.

"Boss, I'm concerned about our pilots. They have been going nonstop now for three months. Everyone is flying almost every day or night, and I'm seeing them making more and more mistakes. This war looks like it is going to continue. I really think we need to take a break."

My gut instincts told me just the opposite.

But he was my operations officer, and I relied on him to have his finger on the pulse of operations. If he felt this strongly, then I needed to pay attention. Still, I wondered why our assessment was so different. Was I facing another blind spot I couldn't see?

I immediately thought, "He must be seeing something I'm missing."

It didn't help that I had been shot down and, as commander, was responsible for doing everything possible to ensure we accomplished our mission while bringing everyone home. If I missed this blind spot, and if my pilots really were as tired and worn out as he was suggesting, would I contribute to getting one of them shot down?

I told him I appreciated his concern and would think about it.

I took a couple of days to reflect on his request.

When I checked in with my boss about taking a day off the Air Tasking Order, which directs military air operations, specifically missions and targets, for twenty-four hours, he said it was my call and I had his support. When I checked with my maintenance chief, he was adamantly opposed to taking a day off. Our squadron was leading the entire campaign for the number of successful launches and recoveries, and it was a major point of pride for the maintainers. When I checked with a few pilots, their reactions were mixed. Some were where I was. Some were fine with a down day. Some felt that flying combat with families at home was taking a toll on the home front.

My gut told me the squadron was doing great and that we needed to stay in the fight.

But I knew I had blinders.

Against my gut instinct, I directed the squadron to take twenty-four hours off the schedule to recharge our batteries for the long fight ahead. We put the day off on the schedule for a week out.

Every time my gut told me I was making a mistake, I suppressed the thought. He's my ops officer, I told myself, and I trust him. I was going to support his request.

Our sister squadron was thrilled at the news. They doubled up on sorties and would surpass us. During one of my flight line walkabouts after announcing the decision, my lead production supervisor, Senior Master Sergeant Andy Anderson, pulled up in his van. Rather than stopping to chat as he did every night, he rolled up the window and drove past looking straight ahead. I knew he was angry with my decision. He is someone I still have great respect for, and it hurt.

It can be lonely at the top.

Combat is not about competing with other squadrons. It's about the mission: beating the enemy and protecting the lives of those entrusted to my care.

As often happens in combat, things change rapidly.

In the week between my fateful decision and our down day, Serbian President Milosevic showed signs that he was willing to stop his campaign of ethnic cleansing and sign a peace treaty.

But the die was cast. We were not on the schedule to fly.

Ten years earlier, I was privileged to be flying on the last day of Operation Desert Storm, the war to drive Saddam Hussein out of Kuwait. As we flew over what became known as the "highway of death," watching defeated Iraqi troops limping back into their country from a free Kuwait, the ceasefire was announced on our radios.

I will never forget that moment. Victory.

I had been part of a successful campaign to free Kuwait from a brutal dictator. I was now the oldest combat veteran in the squadron, leading young pilots and maintainers in their first war.

On the night we stood down for a rest from operations in Kosovo, the ceasefire was announced.

By going against my gut, I denied my pilots and maintainers the experience of victory that I had enjoyed and remembered from a decade before. They had been part of the entire operation from the very first mission and deserved to have the same experience.

It remains one of the decisions I regret most.

In his book *Blink*, Malcolm Gladwell argues that we make better decisions if we can hone our gut instincts and recognize them as we ponder options.

From this experience in command over Kosovo, I worked to better understand my gut instincts and not let others talk me into going against them. As you gain experience in your field, you will develop the ability to understand things instinctively, with reasons that you may struggle to describe in words.

Listening to your gut is a delicate balance, as many leadership traits often are.

Each of us has blinders, and we cannot always see what others are trying to point out because our life experiences are different. As leaders, we must create an environment where we bring folks onto our team who not only see the world through different life experiences but who also feel free to speak truth to power.

Best as we can, we need to listen and be open to other perspectives. But we must also trust our gut instincts. They are formed over a lifetime and come from a central core of who we are, what we have experienced, and how we act when nobody is watching. Such instincts shape our character, and our character is what shapes us as a leader.

As we rise in rank and responsibility in any organization, the decisions get harder. When all the advisers are gone and you are left to make the decision, go back to your initial gut reaction. If you choose to deviate, make sure you can live with the outcome.

When you or others look back at the decisions you made during your time as a leader, the consistency of your decisions should be tied to your character. You might not get them all right. No leader does. But nobody can question your motivation.

Trust your gut.

Chapter 9

Squint With Your Ears

Heather

When I was preparing for confirmation, I knew that the development of space capabilities would be a major issue during my time as Secretary. In my draft opening statement for my hearing, I used the words "space" and "war fighting" in the same sentence. When it was circulated to other government departments for clearance, a holdover at the State Department from the previous administration recommended deleting the sentence. As if by not saying it, it wouldn't be true. I passed word to the State Department that they would have to get someone more senior to tell me to take the sentence out. They didn't, and I confronted the issue directly.

America is the best in the world at space and our adversaries know it. They are seeking to deny us the use of space in crisis or in war.

Once confirmed, I worked with Dave to prioritize developing strategies to meet the threat and capabilities to implement those strategies. In fact, less than twenty-four hours after being sworn in, I was testifying with Dave in front of a Senate subcommittee on space. In response to a question about reorganization and making an independent Space Force, I said that I was more interested in building war-fighting capability than I was in creating a new bureaucracy.

That response didn't sit well with Congressman Mike Rogers of Alabama, who read my comments and was very unhappy. He had reached out to me by phone after I was nominated and talked to me about creating an independent service. I remained skeptical, for a lot of reasons.

A space service would be very small, smaller than the Coast Guard and most state National Guards. It would also be concentrated in a few states, namely Colorado, Florida, Alabama, and California. In practice, over the long term, an independent Space Force might not have effective advocates and could get crushed in the Pentagon and Congressional budget battles once the initial buzz subsided. And, unlike the other services, there was no independent strategic theory of victory from a war-fighting point of view in the same way that strategic bombing had justified an independent Air Force in 1947 and protected its mission.

Moreover, Rogers and his colleague from Tennessee, Jim Cooper, who also was an advocate for an independent service, didn't seem to have a lot of traction in the Congress for their idea, and I knew that I didn't want to spend precious political capital in the early part of my tenure on organizational changes when what we needed most was a workable strategy to meet the threat and the money to develop war-fighting capabilities. The biggest problem wasn't the Air Force organizational chart, and if I focused on that, then I might never get the capabilities the nation needed.

More than most, I also understood the political dynamic on the Hill. Mike Rogers was Subcommittee Chair on the Armed Services Subcommittee that oversaw space and the strategic nuclear deterrent. He aspired to—and eventually did—become Chair of the Armed Services Committee. There's absolutely nothing wrong with that; I aspired to committee leadership when I was in Congress, too. But it did mean he needed to take on a few big issues and raise money for the reelection of his peers. There's not much outside interest in nuclear weapons, but there are about 1,200 space companies in the state of Alabama. He wanted to be a mover and a shaker on space; proposing to make a new military service was a fairly easy way to get attention. I certainly wasn't dismissive of his ideas, but I understood his motivations were not just related to military capability.

So, Dave and I focused on developing space capabilities.

We engaged in wargaming and developed plans to prevail in space through all phases of conflict. Then, we proposed and successfully secured double-digit increases in space budgets to begin to build those capabilities and stripped years out of procurement timelines. We also opened the pathway for rapid procurement from small and innovative companies that weren't suppliers to the Defense Department, and we created the Space Rapid Capabilities Office to pursue high-risk, high-reward capabilities that reported directly to the Chief and Secretary. Throughout, we explained what we were proposing in numerous classified hearings on the Hill.

By 2018, it was clear that Congressional leaders in the House were still focused on organizational structure and wanted to create a new service. And, by this point, their arguments found resonance in the White House.

Having Congressional and Presidential attention meant that we could shape the discussion and not spend our own precious political capital trying to drive some change through the Congress. As long as changes to the organizational structure didn't divert resources from developing war-fighting capability, we decided to alter course and engage the Congress to make sure that a new service was funded sufficiently and set up in a way that would work. In particular, a new service would need to stay under the umbrella of the Department of the Air Force to have top cover and resources, similar to how the Marine Corps is part of the Department of the Navy.

Dave

From 2011 to 2013, I was deployed to the Middle East as the Combined Forces Air and Space Component Commander. In addition to commanding all deployed air forces, I was also the Space Coordinating Authority for US Central Command. My role was to coordinate space support for every US and coalition unit across the Middle East. In meetings with my fellow commanders on their operational plans and space support requirements, it was clear that every single mission relied on space in some way. This was further reinforced when I came back to the Pentagon to serve as the Director of the Joint Staff, responsible for ensuring proper coordination and integration of space across all the services in support of the Joint Chiefs and the Secretary of Defense.

In early 2017, it was time for my annual trip as Chief to the Red Flag exercise at Nellis Air Force Base in southern Nevada. Of all the competing opportunities on my calendar, my trip planners knew this was an event I'd never miss. It gave me the chance to engage face-to-face with our youngest and most talented operators from across every discipline in the Air Force.

When the United States was coming out of Vietnam, a group of visionary leaders determined we were losing too many people during their first ten combat missions. If we could somehow put them through their first ten combat missions in a training environment with exposure to the most intense threats we could muster, then maybe their survival—and therefore our success rates—would climb in the next war. As Airmen, we should be forever grateful for the foresight of General Bob Dixon, Colonel Moody Suter, and several others who created Red Flag.

In the Red Flag auditorium at Nellis, memories flooded over me from exercises I had participated in as a young captain in the F-16 and as director of Operations for Air Combat Command. Even further back, my father, Colonel Bill "Goldie" Goldfein, had also played a part in building this exercise during his time as Chief of Staff of the Weapons Center. To this day, many combat tours later, I remain convinced that there is no better forum to test ourselves in a realistic combat scenario.

The briefing started, and the deployed Air Expeditionary Wing Commander, Colonel Deanna Burt, took the stage. For every exercise, a wing commander is chosen to lead those who have deployed from all over the world to form a joint and coalition team, often labeled the "blue force." Together, they build a game plan to achieve specific objectives handed to them by the Red Flag staff.

Also deployed to participate in the exercise are opposition forces responsible for replicating realistic adversary tactics and procedures to stop the blue force from achieving their objectives. When I was a young fighter pilot in the final years of the Cold War, they often wore Soviet uniforms with Russian nametags and patches.

It is the best training forum on the planet.

As I watched Deanna give clear guidance to the assembled blue team that included allies and partners from several countries, I thought about how far we'd come in the integration of space in every combat mission across the joint and combined force. Watching all the deployed fighter, bomber, airlift, intelligence, special warfare, Soldiers, Sailors, Marines, and coalition teammates report to a space wing commander was refreshing, and it also made perfect sense.

That day at Nellis, Deanna was masterful in her leadership.

These experiences formed the basis of my initial reaction when I learned there was a push from members of Congress to build a separate service devoted to space. In the early days of the discussion, it was led by Congressmen Mike Rogers (R-Alabama) and Jim Cooper (D-Tennessee). Both asserted that space was not advancing as fast as the nation needed, at least not fast enough as a component inside the Air Force. They argued for space as a separate service. There was also discussion about moving it from the Department of the Air Force to the Navy, given the similarities between operating below the surface in submarines and above the atmosphere in space, both "invisible domains."

One of the many lessons I learned from General James Mattis was to do the work up front to clearly articulate a problem statement before acting on a proposed solution. In our initial meetings with both congressmen, it did not appear the problem statement was clearly defined with the requisite data and analysis required to support such a radical change. Nor did there seem to be a shared understanding of the level of "jointness" we had achieved over many years of space development and integration.

Plus, we were actively engaged in a hot fight across the Middle East. How would the Pentagon bureaucracy handle such a monumental change? In separating space from the Air Force, would we inadvertently separate space from the joint fight? Would all the actions required to stand up a separate service speed up and enhance combat capability and readiness or slow it down?

The lack of a clearly articulated problem statement and the potential to lose ground on the jointness central to ongoing combat operations drove my early thinking on establishing a separate service. These were the questions on my mind when asked to testify before the House Armed Services Committee (HASC) on June 21, 2017. As I said in my testimony transcript, "If you're saying the words 'separate' and 'space' in the same sentence, I would offer, you're moving in the wrong direction. That's why the Secretary and I are focused on how we integrate space. Every mission that we perform in the US military is dependent on space."*

In the fall of 2018, I traveled to Maxwell Air Force Base in Montgomery, Alabama, for a series of speaking engagements with our Air University colleges and noncommissioned officer academies. Like my visit to Nellis for the Red Flag exercise, this also provided me an opportunity to hear from Airmen attending school. I was proud to be part of the effort that launched an advanced academic course for space officers. The Schriever Space Scholars Program reflects our best and brightest among majors and lieutenant colonels from across the service. The fellowship enables them to spend a year working toward an advanced degree in space operations under the tutelage of instructors steeped in the business of space.

---

* The Proposal to Establish a United States Space Force: Hearing Before the Committee on Armed Services

I laid out my argument for keeping space inside the Air Force and walked the assembled Schriever scholars and their academic instructors through my experiences as the Air and Space Component Commander in the Middle East, as Director of the Joint Staff, as Vice Chief, and then as Chief. I also shared the discussions I was having in the tank with my fellow Joint Chiefs, the Chairman, and now Secretary Mattis. None were advocating for a separate service or expressing concerns about Air Force stewardship of space matters.

As I spoke with them, there was obvious tension in the room. I could tell they weren't buying what I was selling. After a while, I stopped transmitting and started listening, remembering the saying, "God gave us two ears and only one mouth for a reason."

Active listening is a critical leadership skill we must continually develop the higher we climb in an organization. It is so easy to fall in love with our own ideas and believe with increasing confidence we have it all figured out, but the best leaders never stop growing. Being able to listen well, or "squint with our ears," and truly hear what others are saying is one of the most important tools in a leader's tool bag. It was time for me to sit back and listen to my Airmen, even if—especially if—their ideas differed from mine.

One of the senior civilian instructors expressed frustration that an article he penned advocating for a separate service had been stopped before its publication. "Do we not have academic freedom to express our opinions here at Air University?," he asked. I hadn't heard about it, so told him I would investigate the matter immediately and get back with a personal answer. Academic freedom is an essential element at any college, and I wanted Air University to be a champion of rigorous debate. It took only a couple days to resolve this and get his paper published.

As the conversation continued, I could tell each of the Schriever scholars had a lot to say. Every opportunity I had to get ground truth in small forums was liquid gold to me as a leader. They shared what it felt like to be in a service dominated by air operations where space was considered second tier. They recounted story after story of programs that could have advanced space operations being squandered or canceled as they competed with other procurement programs like the F-35 or B-21. After over two hours of discussion where I said very little and took many notes, I closed the session by thanking each of them for speaking truth to power. These were my Airmen, and they convinced me that I had developed my opinions based on experiences as a joint leader but had not taken the time to understand perspectives of other stakeholders in the debate.

Upon returning to the Pentagon, I told my travel team I needed to do a "listening tour" across the space enterprise to include our bases, civilian agencies, commercial companies, and our allies and partners who were wrestling with similar issues.

I traveled to every base engaged in space launch, operations, command and control, and acquisition to meet with Airmen and hear their stories. I visited companies like SpaceX, Boeing, and Northrop Grumman to meet with industry leaders and walk their production lines. I accompanied Congressman Jim Bridenstine (future NASA director) on a visit to Cape Canaveral to witness a SpaceX launch that included the first-ever recovery of the first stage of the rocket booster on land. On overseas trips, I met with international air chiefs to better understand how they were thinking about the space domain and what organizational construct would benefit them most since they usually followed our lead. Throughout, I relied heavily on two senior space officers I had known for most of my career and whom I respected deeply. Generals Jay Raymond and John Hyten had both commanded space units at every level. Their wise counsel was formative.

For the next several months, I compiled a list of questions to help me come to a more thoughtful conclusion about whether to support the establishment of a separate service for space. In the end, I whittled it down to three major questions:

1. Could the Air Force embrace space superiority with the same level of passion that we currently apply to air superiority?

2. Could I, as Chief of a service with an expansive list of responsibilities that spanned every domain from 100 feet underground in a nuclear missile silo to the outer reaches of space (and everything in between), advance joint space operations for the nation as fast as a Joint Chief singularly focused on space?

3. Given the combination of two technological advancements (much smaller but highly capable digital payloads and significantly reduced launch costs), could I optimize and harness the rapidly expanding commercial sector as efficiently and effectively as a Chief of Space Operations?

While on my listening tour, the conversation was gaining traction in Washington, including at the White House. President Trump started talking about a separate service and put Vice President Pence on the issue. This was good news for me and Secretary Wilson, since we both had a pre-existing relationship with the Vice President. Heather had served with him in Congress when he was the representative from Indiana. I had interacted with him early in his tenure as Vice President in several areas involving the Air Force. His rebuilding of the National Space Council was an important part of the journey. I always found him to be thoughtful and engaging and was glad to be working with him on this important issue.

Heather

By June 2018, it was clear that the President was enamored with the idea of an independent Space Force. It didn't matter why, and I don't think it was because he was unhappy with what we were doing with space capability. In fact, during his first term, space was a priority and a lot got done.

Every political appointee in any administration needs to understand that there will be some issues that the President chooses to elevate to their own level that might otherwise be delegated to you. You might be asked for your advice; you might not. The President and Vice President were elected. You weren't. And the reality is, those issues the President decides to take an interest in have a lot more likelihood of getting done than something advocated for by a Service Secretary and a couple of House members.

President Trump liked the idea of an independent Space Force. He announced his intention to create it, and, once that decision was made, our responsibility as Secretary and Chief of Staff of the United States Air Force was to make it a reality in a way that would work for the long haul.

Dave

After several months of listening to and engaging with as many stakeholders as I could find, I determined the answers to all my questions were "no."

Changing organizational culture is among the hardest tasks for leaders of large organizations. You don't just order it from above. There must be bottom-up buy-in, from the squadron level all the way to the top. I felt strongly that a shift in our culture to embrace space superiority could be achieved, but not in a short amount of time. Still, we needed to move quickly to stay ahead of the threat.

I also determined that a service chief singularly focused on space could advance the military/civilian/commercial ecosystem faster than I could with everything else in our portfolio of responsibilities, including an ongoing hot fight in the Middle East that was not going well. The combination of significantly reduced launch costs plus miniaturization of satellites opened space to new missions and new participants. What was once the purview of only large nation-states with billions of dollars to spend was now open to small start-ups and developing countries across the globe. A Chief of Space Operations could harness this commercial, civil, and international expansion more effectively and efficiently than I could.

Adding these together, I decided a Chief of Space Operations could more effectively lead and support the joint team in his/her role as a member of the Joint Chiefs more effectively than I could leading both air and space operations.

So, I changed my mind and became one of the biggest proponents of a new service.

In a discussion with Secretary Mattis and the Vice President and President, I shared my evolution in thinking that resulted from the listening tour. My problem statement centered around the rapid expansion of military, commercial, and civilian space activity both in friendly lanes and among potential adversaries. A service and joint chief singularly focused on space could best harness the rapid advancements taking place. My advice to Secretary Mattis was to keep the Space Force inside the Department of the Air Force in the same way the Marine Corps and the Navy coexisted inside the Department of the Navy. He was in complete agreement.

In March of 2019, Heather was offered a position as President of the University of Texas–El Paso (UTEP) and Secretary Mattis had just left the Pentagon. Having previously served as President of the South Dakota School of Mines and Technology, Heather knew she wanted to return to public higher education in the west after her time was up with the Air Force.

There is no fixed tour length for a Service Secretary, and when she shared with me the possibility of taking this opportunity, I encouraged her to go for it. I hated losing her, but I knew she would make a positive difference for students and higher education in the same way she made a difference in the lives of Airmen across the service.

President Trump soon announced Barbara Barrett as his choice to succeed her as Secretary, and I began working with her to advance the idea of a new service with colleagues on the Hill and in the White House.

Barbara was the perfect follow-on to Heather. While people didn't know it at the time, Heather worked the phones to interest talented people in the Secretary job so that President Trump had a list of good candidates. Heather happened to know Barbara well, and Barbara was clearly the right person at the right time for the Air Force with her background in both space and diplomacy.

Among the many elements of her impressive resume, Barbara trained in Russia to be the backup astronaut for a launch to the International Space Station in 2009. She was also a former US Ambassador to Finland and former Deputy Director of the Federal Aviation Administration. As if it couldn't get any better, she is a gifted speaker who squints with her ears, quickly captures the essence of an issue, and is a natural visionary. I was incredibly blessed to work with these two leaders on this important journey.

Once we knew we were headed toward creating a new service—the first new service since the Air Force was created in 1947—we needed to select the right leader to take on this extraordinary task. I had known Jay Raymond and his wife, Mollie, for many years. Whenever I had any question related to space, Jay was my first call. He was operationally savvy, politically astute, and among the finest officers I have ever worked with. His character was also beyond reproach, which was exactly what this new service needed.

Anyone in the military knows it is a team sport when it comes to family, and this is especially true when it comes to service chiefs and their spouses. Dawn had taken on so many responsibilities as the First Lady of the

Air Force, and I wanted to make sure Mollie fully embraced the chief's role and was on board with what could become hers. Our families had been friends for years, and she is a wonderful wife, mom, and leader. Like Jay, Dawn and I knew she would be perfect.

I called Jay into my office. "Jay, I know you were recently confirmed as the combatant commander for space," I said. "I have another option for you to consider." We discussed him becoming the first Chief of Space Operations—the equivalent of Hap Arnold, Billy Mitchell, and others credited with starting the US Air Force. I encouraged him to take some time to discuss the opportunity with Mollie and get back to me in a few days.

He came to see me the next day. "Mollie and I are all in."

We shook hands and never looked back. Secretary Barrett agreed with my recommendation after conducting her own interview and took Jay's name to Secretary of Defense Mark Esper. It was not long before both the Vice President and President both heartily endorsed Jay. Because he had been unanimously confirmed as the US Space Command Commander, he served in both roles for a year before they were split, and Jay and Mollie were able to devote 100 percent of their time building the newest service.

As leaders, we must be willing to listen to those around us and adjust course when circumstances change or when we gather information that challenges our initial assumptions. It is not a failure to admit when we see the need to move left when we initially believed the correct direction was right. Failure occurs when we let ego, arrogance, or insecurity drive us in a direction knowing in our heart we're going the wrong way. The best leaders continue to develop listening skills as one of the most important tools in a leadership tool bag.

Squint with your ears.

Chapter 10

Shovel the Snow

Heather

The South Dakota School of Mines and Technology is a forty-minute drive from Mount Rushmore, just east of downtown Rapid City, South Dakota.

I had been there as president only a few days when someone described Rapid City as "the banana belt of South Dakota."

I wasn't buying it.

While the beautiful Black Hills of western South Dakota provide some protection from the extremes of wind and weather experienced by the eastern part of the state, there weren't any banana trees.

I arrived and started as the university's president in July 2013. There are few better places to hike and bike and fly fish than the Black Hills in the summer and fall, and I took advantage of it.

The weather in the last week of September was gorgeous. Blue skies, calm winds, and temperatures in the seventies had students lounging on the central quad in shorts and T-shirts.

So, when faculty member Darren Clabo—who was also the state's fire meteorologist—sent me an email saying, "I know this sounds weird, but all my meteorological models are predicting we will have forty inches of snow on Saturday," my response was understandable.

"You're kidding me, right?"

He wasn't.

It was going to be a busy week. The governor, the chancellor, several regents, and special guests were arriving on Thursday for a very large scholarship dinner. I was to be inaugurated as president of the university on Friday morning under the ancient trees on the central quad.

As the days progressed, local weather reports started converging with Darren's models. A terrible storm was indeed heading for South Dakota.

On Thursday evening, we drove to the VIP reception before the scholarship dinner in downtown Rapid City as the heavens opened with torrential rain. I got the latest weather forecasts from Darren and, at the reception, huddled with the governor and regents. We'd been planning this event for months but needed to make a change. We decided to have the dinner quickly, dispense with long speeches, swear me in at the podium as president, then tell people to go home and hunker down. We also canceled the inauguration festivities the following day.

The governor and our special guests headed to the airport to beat the storm. My husband, Jay, and I arrived at the president's residence as the rain started turning into wet snow.

It snowed all day Friday and winds whipped at fifty to seventy miles an hour. By Saturday morning, the snow had stopped. As I looked out the kitchen window down the slope to the valley, it didn't look too bad.

Then I opened the garage door. Now, I grew up in rural New Hampshire and have seen some high snow. But I had never seen anything like this—and sure as heck not in October. Over forty inches of wet snow blanketed everything. Drifts reached the eaves.

Jay and I put on layers of winter gear and took out the huge red snowblower. It had been a long time since I'd seen one of these in action. We quickly discovered the first rule of snow blowers: they must be taller than the snow. They don't tunnel worth a damn.

We hadn't lost power at our home, though many others did, which allowed us to communicate with the dean of students and the housing staff to check on the status of the thousand or so students in their dorms. It turned out they had power and there were enough student employees to keep the cafeteria running. Only one slip-and-fall injury was reported, and one faculty member from the metallurgy department who stayed late to grade papers on Thursday night got stuck. He was a nice guy, and the students liked him, so they fed him, too.

But the roads were completely blocked and the whole of Rapid City remained at a standstill.

By noontime on Sunday, the sunshine had melted away a lot of the snow and the main roads were mostly cleared. If we could get our four-wheel-drive vehicle out of our neighborhood, we thought we could make it safely to campus. We knew we likely wouldn't make it back up the hill to the residence, so we loaded gear in the back of the car and, with a few pushes from helpful neighbors, got out of our cul-de-sac and headed that way.

None of the parking lots or sidewalks at South Dakota Mines were cleared. We parked on a side street and hiked in through the snow, dragging our shovels. As I rounded the corner of the classroom building that looks onto the central quad, I froze. All the beautiful trees that shaded students lounging in shorts and T-shirts only a few days before had fallen. The rain turned to ice and the heavy, wet snow toppled them.

In total, we lost seventy trees to the storm and only one building experienced minor damage.

Jay and I circled the campus and headed for the student union. We squeezed through the main door, partially freed from snow and ice blown against it by the wind, and found students, generally cheerful (as they usually were), playing games and enjoying each other's company.

The students at Mines are all engineers and scientists. Most of them have grown up in the upper Midwest on farms and ranches and in small towns. They are hardworking, head smart, and hand smart. By now they had a bit of cabin fever in the dorms, but everyone seemed fine.

Jay and I went back to the entrance of the union and began shoveling snow away from the door so that people could exit quickly if needed and to prevent more slips on the packed snow that was turning to ice.

As we shoveled, a young man in a baseball cap and sweatshirt came out of the union and asked with a big smile if we needed a hand.

"That would be great!" I said. "Do you have a shovel?"

"I think I know where I can get one," he said.

About ten minutes later, a pickup truck arrived with a half-dozen young men in the back carrying shovels they borrowed from one of the fraternity houses. The students jumped out, greeted us warmly, and helped us clear the path.

By the time the sun started to set, we had the union entrance and sidewalk cleared.

"You guys hungry?" I asked. (I can never recall being told "no" in answer to that question while at South Dakota Mines.)

We went into the union, downstairs to the cafeteria, put our wet outer clothes and mittens on the chairs around a big table, and got in line for some chicken fried steak and mashed potatoes with gravy. I have to say, it really tasted good.

As we ate dinner together with the students, the young man with the ball cap who had organized the shoveling brigade turned to me.

"President Wilson, can I ask you something?"

"Sure," I said.

"Why were you shoveling the snow? It doesn't seem like something the president of the university ought to be doing."

I smiled at him from under my steaming knitted watch cap.

"It needed shoveling," I said.

He grinned back. "Yeah, I guess so."

That night, Jay and I stayed in a hotel downtown a few blocks from the university. School would be closed on Monday, but we sent out a message to all the faculty, staff, and students that anyone who could safely come to campus should bring their tools and help clear the campus. Hundreds did.

With the aid of the Theta Tau fraternity that brought chainsaws to cut trees and the football team that worked all day with head coach Stacy Collins, along with dozens of faculty, staff, and students, we had campus ready for class on Tuesday.

The blizzard of October 2013 was a disaster for western South Dakota. Tens of thousands of cattle still in summer pasture got wet with rain and then froze to death or suffocated in the drifts when the temperature plummeted.

As a leader, particularly when something goes terribly wrong, there is no substitute for showing up and participating in the work that must be done. You can't just walk by and encourage people. You can't just monitor the situation from afar or give orders over a radio. You can't be too good to do the heavy lifting.

Sometimes, the most important thing you do is participate in the work at hand alongside others. They will respect you for it. Many will follow your lead and join you in the work, which will get done sooner as a result.

When the snow had finally melted and things returned to normal, Darren Clabo gave me a framed picture of his weather model of the Black Hills for that day. The cabinet also gave me a new name plate for my desk.

Except it didn't say "University President" or "Doctor Wilson."

It said, "You're kidding me, right?!"

I've kept that name plate on my office desk ever since. When you're a leader, there will be a lot of times when someone walks in and says, "Um, boss, there's something you need to know." When they do, sometimes the only appropriate response is, "You're kidding me, right?"

When the reality of the situation becomes apparent, do what needs to be done—and do so cheerfully. Others will take notice. They'll probably even join you.

Shovel the snow.

Chapter 11

Step Up and Lead

Dave

"Have a good day, girls." I kissed Dani and Diana as they ate their cereal and got ready for school. Our daughters attended St. James Elementary School in nearby Falls Church, Virginia, where Dawn taught third grade. After a hug from Dawn, I headed out to my car.

We had the perfect setup for my commute to the Pentagon where I worked as a budget programmer in the Combat Forces Division. My boss, Colonel Blair Hanson, jokingly called us the "breaker of dreams." He'd remind us often that "others design the Air Force we need" but that our job was "to build the Air Force we can afford."

My commute was a dream compared to so many of my colleagues who had to join strangers to ride in the faster HOV lanes (affectionately known as "slugging") or catch buses or trains every morning. By contrast, my twenty-minute drive was a nice start to the day as I caught up on the news and thought about how to optimize every tax dollar we were given by the American people. We never forgot that every penny we spent did not belong to the Air Force. It belonged to American citizens who worked hard, paid taxes, and deserved to be defended by the greatest Air Force in the world. Our job was to find the best options for investing their money with appropriate oversight by elected leaders in Congress.

It was an especially crisp fall morning with the sky a beautiful blue. I normally drove in on my own but decided to join fellow Pentagon commuters on the bus traveling north on I-95. As I crested the hill, our majestic Air Force Memorial filled the windshield with its three towering spires reaching toward the heavens. The date was September 11. None of us had any idea that aircraft in the sky above would be taken over by terrorists who would change our lives forever.

At the Pentagon, I made my way to the small bar we built in our secure office suite. Most of us had come from the fighter and bomber business and enjoyed gathering there at the end of each day. The Pentagon had been undergoing the first major renovation since construction in January 1943. Corridor by corridor, the entire building was being gutted and reinforced with shiny new floors, wider hallways, shatterproof windows, and a historic art collection that matched any in DC. Our corridor faced south and was the last one scheduled to be renovated, so we were enjoying a few more months before it was our turn to move.

"Good morning, Fingers," Blair offered as we both went for the coffee pot. "Thanks for taking the meeting for me." Blair was scheduled to attend a meeting in Rosslyn later that morning and asked me to attend in his place. As his deputy, the only acceptable answer was, "Sure, boss." It was the first time I was to attend a meeting outside of the building, and I was not looking forward to a bus commute. We stood at the bar with our steaming mugs then looked up at the news. The first aircraft hit the World Trade Center.

Like most of the world, we thought it was either an accident or some kind of explosion inside the building. We watched quietly, not understanding we were under attack and that another aircraft was headed our way.

But it was time to head to the bus stop if I was going to make the meeting. I finished my second cup of the day and left.

During the short ride, someone on the bus informed us that the other tower had been hit. In 2001, cellular phone technology was very limited. This person found out by talking to someone in front of a television.

Not long after arriving in Rosslyn, I heard someone shout, "Oh my God!" and I looked up in time to see an American Airlines tail disappear behind the buildings followed by a loud explosion with a plume of billowing gray and black smoke coming from the direction of the Pentagon. When we saw the second plane hit, we knew what was happening was no accident. But watching it unfold only a stone's throw from my office, where less than an hour ago I was enjoying a hot pot of coffee like any typical morning, we all knew America was under attack. Everyone poured into the streets and stared at the sky as the serene, crystal blue was replaced by a cloud of death and destruction.

I needed to get back to my team.

I ran to the subway and jumped on the first Metro train headed back to the Pentagon. Rosslyn was only two stops away on the Blue Line, yet unbeknownst to anyone, an emergency switch had been pulled directing every train out of the city, so I was heading in the wrong direction. I contemplated my options as the train headed westward to Falls Church. Cell phone coverage was down. Passengers exchanged worried looks.

Somehow, my brother Mike got through to me with the latest as I made my way toward Dawn and our daughters at St. James Elementary. He shared that two airliners had hit the Twin Towers and one hit the Pentagon. It was a Godsend that we connected, since he was a pilot with American Airlines and was on call that day to fly while I was supposed to be in the Pentagon.

All this time, Dawn was living a nightmare.

In her world, it was also a normal Tuesday. A little after 10 a.m., she went to the front office to make some copies for her class. That was when she learned the news. The secretary looked up and told her the Pentagon was just bombed. Dawn tried her phone. No luck. Cell phone towers were overwhelmed with calls.

Back in her classroom, distressed parents started arriving to pick up their children. As her class emptied out, a young boy remarked, "Mrs. Goldfein, there sure are a lot of appointments today." Many of the remaining children in her class had parents who worked at the Pentagon. Thoughts kept turning over in her mind: Have I just lost my husband in another attack after he survived an enemy missile over Serbia? Like so many others, our attempts to call each other were in vain. We were left waiting. Wondering. Worrying.

An hour later, Dawn went out in the hallway to try her phone again. As she looked up, I walked toward her in my blue uniform. She let out a scream of relief as we ran and embraced. We had no idea this day would define the next two decades of our lives.

After arranging her a ride home, I took her car and drove back toward the Pentagon. As soon as I left, Dawn went to Diana and Dani's classrooms and told the girls that Daddy was okay. Dawn understood I had to get back to my team. The smoke across the horizon still hung in the air.

Every entrance to the Pentagon was blocked. I kept calling Blair and members of the team. Silence. The radio announcer mentioned a team was being assembled nearby at Fort Belvoir to fly into the Pentagon and offer support and medical assistance. At that point I turned south, drove to the hospital, and signed up for the first team going in. An hour passed. No movements. I got back into the car to try once more to get into the Pentagon, all the while still trying to get ahold of the team. Yet again, no luck.

Circling and looking for an opening, I noticed a sign for the Red Cross headquarters. They would be going in for sure, I thought. When I introduced myself and asked to be placed on the first team, I was told it was full but that I could get on a later team the next day. I pulled out my Pentagon credentials and asked if they had anyone on the first team with access to every part of the Pentagon. They didn't. Next thing I knew, I was given a Red Cross T-shirt and headed to the waiting bus.

We arrived at the rescue effort with scores of firefighters, police, and medical first responders all working feverishly to find survivors. By now the sun was starting to set on this tragic day. Our job was to help with anything that was needed. We found the police and fire chiefs and introduced ourselves, asking what we could do to help. "Right now, the heat is so bad we need water for the guys coming out of the building." So, we set up a line of volunteers and stationed ourselves as close to the building as possible, handing out water to everyone coming out. Others were pulling out survivors.

There was one scene forever etched into my memory. As we kept the water line going, someone pointed up at the deep crack in the west side of the building where the plane struck, flames and smoke still billowing. On the very edge of the opening, on the fourth floor, a lone computer sat untouched on a desktop despite total chaos and destruction around it. I stared at it for several minutes. I couldn't put it to words then why I was so struck by the sight of it, but looking back now, it's clear to me that it represented both randomness and pockets of order inside that fateful day.

Much of life and history plays out in the wake of near misses. My brother, an American Airlines pilot, was on call instead of in an airport or in the skies. I was supposed to be in the Pentagon that morning but left the building to attend a meeting in Rosslyn for a colleague. The al-Qaeda murderers planned to hit the east side of the building where senior defense leaders were at work, near where my team was located. Instead, they missed and circled around the Pentagon before hitting the west side, which had been recently renovated with reinforced walls and windows. And the so-called pilots misjudged their glidepath for a second time and hit the highway before bouncing into the Pentagon, causing the energy to dissipate enough to stop it at the third corridor.

While massive loss of life was spared, 184 people died in the Pentagon attack, including the passengers and crew of American Airlines Flight 77. My thoughts and prayers go out to everyone who lost a loved one on that tragic day.

As the Red Cross team continued to support the rescue effort, I noticed a lot of activity in the south parking lot adjacent to our work area. Wandering over, I met a team from the Salvation Army. They asked if I knew who was in charge, and I pointed them to the police chief I spoke with earlier.

A steady flow of trucks and vehicles showed up to offer support for the effort. Food, medical equipment, supplies—all kinds of services and support kept coming in without letup. It was both our worst and our best day as a nation. We had been brutally attacked. And yet, it is in these times that the true spirit of America shines. People from across the country dropped everything to offer love and support to strangers. We may have our philosophical differences, but when attacked, no community comes together like the American people. It should make every known or potential adversary pause and think. Don't wake the bear. In peacetime, we air our differences noisily and proudly as the founding fathers intended. We appear to some as divided and to others as chaotic. But attack us, and you will face the most united and disciplined nation on the planet.

The events and aftermath of September 11 reinforced the decision I made years before to return to the Academy. America deserves to be defended.

Back in the chaos of the south parking lot, I thought back to the three basic rules of leadership I had learned from General Mike Ryan: put someone in charge, build a plan, and follow up.

Someone needed to step up and lead.

I looked over and saw Sheila Mulhern, a good friend who also worked in the Pentagon. Sheila was a civilian in the personnel section on the Air Staff, and I knew her as a great leader and go-getter. She had the same instincts kick in. "Someone needs to take charge here," she said. "Look at all the trucks coming in."

I was able to get another Red Cross worker with credentials to drive me back to our house where I could see Dawn and the girls for a few moments before returning to the support effort in uniform. After walking back into the south parking lot, I introduced myself to those assembled and declared that Sheila and I were in charge of the support effort. Nobody questioned our authority. The police and fire chiefs nodded appreciatively as I re-introduced myself as Colonel Goldfein and offered to lead the supporting effort.

Sheila had already commandeered a tent to serve as our makeshift headquarters. With a whiteboard, we listed the agencies as they arrived and grouped them under two headings: Red Cross and Salvation Army. Anyone who showed up with money or services went to the Red Cross; anyone who showed up with supplies and equipment went to the Salvation Army. It was rudimentary, but it worked. Sheila came up with the name "Camp Unity." We spent the entire first night organizing the effort as it grew.

I marveled at the response. We needed fencing and lighting, given 24-7 operations. A truck from Home Depot arrived with everything necessary, including a team to assemble it all. The next day, three women asked if they could help. "What can you do?" I asked. They said they closed their hair salon to come help. We put them to work under the Red Cross banner and soon enough they built a makeshift parlor for the responders. Folks showed up from all over America. On the afternoon of the third day, an eighteen-wheeler pulled in. "We're from a men's Baptist group in Georgia," they shared. "We're headed to New York if you can't use us here." I looked at the huge truck. "What can you do?" I asked. "We can make gumbo," he offered. He chuckled when I responded that it would be a lot of gumbo, then went on to describe how they could serve it out of a circus-like tent, that they had everything needed to take care of this size of a crowd or more. Sheila and I looked at each other. "Why not?" we agreed.

Today, if you see pictures of the south parking lot after 9/11, the large white tents are clearly visible in the center. These patriots fed the entire rescue, recovery, and support teams for weeks, all at their own expense. Camp Unity grew into a support area that included tents for every business or group that offered support. Nobody was turned away. Whenever we needed help, either the Red Cross or the Salvation Army embraced and organized every volunteer effort.

I encourage you to take a step back and think about this. America was brutally attacked without warning, with thousands dying in a horrific way. Rather than hunkering down and focusing only on themselves or their families, these everyday Americans stopped whatever they were doing to race into the breach to help their fellow citizens.

Where else but America? I love this country.

Two days after we were up and running, I was told a congressional delegation would be visiting. Sheila and I arranged for several of the lead agencies to walk the delegation around as we stood in the back. I didn't know it then, but one of the visiting leaders was my USAFA classmate and future Secretary of the Air Force, Dr. Heather Wilson. This was one of many moments our paths crossed on our individual journeys to servant leadership.

We took Camp Unity down after about three weeks and transitioned our work to supporting the global "war on terror." A month later, when I ran past the Pentagon during the Marine Corps Marathon, I could still see those white tents in my mind's eye. But I also noticed something else still on scene: the lone computer on the edge of the crevice sitting defiantly in its place. These remnants of loss, whether memories or physical relics, serve as important reminders of how much we suffered and endured. But they also represent the true strength of America: an America that was tested and tried but rose above as one body, one nation under God, indivisible, and defiant against those who would do us harm.

There are times when you will find yourself in situations of crisis, tragedy, loss, or chaos. Sometimes situations will entail all these things at once. Such circumstances are usually in dire need of strong leadership. When that moment arrives, step up and lead. Others will be better for it, and so will your country and our world.

Step up and lead.

Chapter 12

Do What You Can, From Where You Are, With What You've Got

Heather

The sky over Washington, DC, so often depressingly cloudy for a Westerner, was crystal clear blue that September morning. I was between apartments and sleeping on my office couch in the Cannon House Office Building, so I got up early, went to the House gym to work out, and was at my desk long before the staff arrived.

My husband, Jay, a member of the New Mexico Air Guard, had boarded a flight to Washington, DC, from New Mexico to attend a meeting planned for September 12 at the Pentagon. I rarely got to enjoy his company in Washington and was looking forward to having dinner and spending the night in a hotel with him.

I was sitting at the round table in my office with Jim Chavez, my science fellow from Sandia National Labs, talking about what he might do for the last six months of his fellowship to make it most meaningful, when there was a soft knock on the office door and one of my young staff members came in.

"Congresswoman, a plane has hit a building in New York," he said, reaching up to turn on the television high on the wall of my office. Several staff members came into my office or huddled around the television in the adjacent staff office across the entryway, watching the news.

It was a little before 9 a.m.

While a lot of military members and intelligence officers have said in the years since that they immediately knew that this was a terrorist attack when the first plane hit, I did not. I had served on the Intelligence Committee in the House and was, at that time, a member of the House Armed Services Committee. But the "failure of imagination" that allowed terrorists to use a civilian airliner as a weapon affected me too. I thought this was some terrible accident.

And then, at 9:03 a.m., my staff and I watched a second plane hit the south tower of the World Trade Center in New York.

This wasn't an accident.

We continued to watch the footage and then, about half an hour later, there was news that there was smoke over the Pentagon.

It wasn't just New York.

When bad things happen, you start by counting noses. You begin to do what you can, from where you are. We got the office together. "Is there anyone not here today?" I asked.

Capitol Hill staff are generally pretty young, usually right out of college. As it happened, the youngest and newest member of our team, an intern, had gone downtown to get a report on unsolicited commercial email that was supposed to be coming out that morning. Everyone else was accounted for. Someone quickly confirmed by phone that our intern was okay and told him to return home rather than come back to Capitol Hill.

Telephone lines were jammed, but I told my staff to make sure they kept trying to call their families to let them know that they were okay. We knew the bridges over the Potomac were likely jammed too, particularly near the Pentagon. So I asked them to sort out who lived in DC and had extra room to take in others who might not be able to get home.

Our staff member from Long Island, New York, clearly shaken, asked what would happen now.

Before I could answer, Barbara Cohen spoke up. Barbara was my executive assistant who managed the chaos of my schedule. A single mother of two teenage girls, very active in her church and a native of Washington, DC, Barbara was usually quiet in meetings.

With the steady voice of a woman who had seen her share of hardship, Barbara said, "We carry on. We take care of the people who need help, and we get up tomorrow and go to work."

I left the staff to sort out housing arrangements and returned to my office. All aircraft in US airspace were being diverted to land. I needed to get in touch with Jay. While he was airborne during the attack, for some reason, I was never worried about him. I knew from his departure time that his aircraft wasn't near the East Coast, so the aircraft already down were not his. Still, while we knew very little about what was happening on other aircraft in the sky, somehow, I just knew he would be fine.

I was able to leave a message on his cell phone. His aircraft had been diverted to El Paso, and he called me as soon as he was on the tarmac. I gave him what information I could and told him to head to the rental car desk and drive home. Aircraft wouldn't be flying anytime soon. Within an hour, he and three strangers shared a car and started driving back to Albuquerque.

I was also concerned about our children. Our son, Joshua, was in second grade. He knew his mom was in Washington and his dad was flying to the Pentagon that morning—not a good combination considering the morning's events.

After multiple tries, I was able to get through to our longtime babysitter and told her, "Call the school principal and ask her to go to Joshua's classroom. Ask her to tell him this: Your mom called. She's okay, and your dad is driving home tonight for supper." While I always made it a point to be just "Josh's mom" at our neighborhood public school, the principal and his teachers knew my role in the community and where I was during the week. I knew the principal would get that simple message through and that it would reassure Joshua if he got snippets of troubling information during the day.

I also told our babysitter to keep the television off. I'm glad I made that decision.

I went out to the staff room again in the Cannon Building.

By then the House session had been canceled for the day. My team had sorted out the housing needs, most had been able to get through to their families to reassure them, and I sent them home. Jim Chavez chose to stay with me in the Cannon Building.

While extensive continuity of government plans were in place for the executive branch, at that time, there really wasn't a plan for the Congress. In fact, once the Capitol building itself had been evacuated, communication was very limited. Some of us had Blackberries, but the communication systems to offices were operated from the Capitol building itself, and no one was there.

Once the staff had gone, I locked the main door and sent an email to the staff in Albuquerque, letting them know that I was remaining at the office with Jim, the DC staff was accounted for, and they were going home.

I waited a few minutes in the silence of my office and then resent the message asking someone to confirm that they had gotten my message. They did. All accounted for.

I also started trying to contact a friend from New Mexico, Kip Nicely, who was a Navy officer assigned at the Pentagon. I was eventually able to get through to him. He was okay, but clearly shaken. That morning, he had just left the building to go to a meeting at the Pentagon annex when Flight 77 flew over his head and slammed into the side of the Pentagon where his office was. He told me they were setting up operations near the building. He asked me if I would come over. I told him I would try to get there.

Less than an hour later, while I continued to listen to the news, a knock came at the door. A Capitol policeman said that they were evacuating all the buildings. Jim and I closed up and initially started walking to Jim's apartment when a Capitol policeman recognized me as a member of Congress. He told me they were sending members to the Capitol Police Station on the Senate side of the complex.

We turned and started to walk there. The sky was still intensely blue as we walked down the middle of First Street, bereft of any traffic, in front of the Supreme Court, with a law enforcement officer stationed every forty feet or so keeping people away from the Capitol. It would be some time before we learned that, were it not for the passengers on Flight 93 that crashed in Pennsylvania, the scene of quiet would have been quite different.

As we reached the Capitol Police Station, members were allowed but not staff. I thanked Jim for sticking with me that morning, sent him home, and entered the police station. On one of the upper floors forty or fifty members of Congress were gathered. There were some chairs, a television screen, and a few landline phones. No members of the Congressional leadership were there, which became a problem as the day went on.

I had tried to contact my mom several times without success. Cell service in Washington was still overwhelmed. I decided to try the landline and dialed a number in rural New Hampshire. My stepfather answered.

Joe had not been my dad growing up. He was a simple man and was caring toward my mom.

"Joe," I said. "Is Marty there?"

"No. Took the dog somewhere. I don't know."

"Joe, when she gets back, I need you to give her a message."

"A-yuh."

"Tell her, 'Heather called and she's okay.' "

"A-yuh. . . . About that Social Security thing. You going to get that fixed?"

It's possible he didn't have the television on. Even so, everyone sees the most important priority for the Congress through their own worries.

"Joe, we'll take care of Social Security. I promise. But when mom gets home, just tell her I called and that I'm okay."

"A-yuh. Took the dog somewhere."

"Okay. I'll try to call again soon. Okay?"

"A-yuh."

I hung up and laughed to myself. You can't make this up.

Every few hours at the police station the House Sergeant at Arms came to give us an update, with the leadership on a conference call listening in. But he really didn't know much; we were finding out more from the television than we were from the official briefings. That created frustration that built in the afternoon.

Porter Goss was the chair of the House Intelligence Committee. I had noticed that he wasn't in the police station and suspected that he might have evacuated with Speaker Dennis Hastert, who, third in line for succession in the government, would have been taken to a secure location as part of the continuity of government plan. A former CIA agent, Porter was close to the Speaker, and he and I had worked together on the House Intelligence Committee. I knew he would trust my judgment and could communicate directly with the Speaker, even if they weren't co-located.

I stepped outside the Capitol Police Station and called him to let him know that frustration was building among members, which may not have been entirely clear listening on the conference call. It was probably important for the leadership to return to Capitol Hill, if they possibly could.

When it got to the end of the school day in New Mexico, my phone rang. It was our babysitter. She had just picked up Josh and he wanted to talk to me. Our family rule was, no matter what, they could call me any time.

Even in second grade, Joshua was a serious kid. He wanted to understand things.

"Mom, what happened?" he asked.

I told him what I could, in terms a second grader could understand, and sought to reassure him.

But sometimes, children ask the most important questions with a clarity that we lack as adults.

He paused after my explanation and then he asked, "Why did they do this to us?"

The hardest questions are the simplest ones.

"Bud, evil exists. What matters is what we do in the face of evil. We'll do the right thing. We'll protect people. We'll choose to do good in the face of evil."

"When will you be home?" he asked.

"I don't know, Bud. But I'm fine and Daddy will be home by suppertime," I said.

"Love you buckets," I said.

"Love you too," he replied.

The briefing at around 5 p.m. was tense. The leadership was still on the conference call line. There wasn't much more information on the security situation, but members made clear that they believed we needed to stop hiding. The American people needed to see the government would continue and that we would not be cowed by terrorists. Of course, the Capitol Police and Sergeant at Arms cautioned about the security risks.

In Republican Conference meetings, I was usually quiet unless I had something really important to say. A few people had spoken about the need for the American people to be reassured, and the conversation didn't seem to be going much of anywhere.

My colleague Doug Ose of California was sitting in front of me and could be counted upon to be direct. He stood to speak and was loud enough for those on the conference call phone to hear him clearly.

"I may be slitting my own throat by saying this," he said. "But I think you people in leadership are a bunch of goddamn cowards. We'll be gathering on the East front of the Capitol at 7 p.m. tonight to reassure the country whether you show up or not."

The leadership didn't commit themselves one way or the other, but they knew where things stood.

Daniel Coughlin, a Catholic priest and Chaplain of the House, had joined us earlier in the afternoon. After the briefing drew to a close, he offered a prayer and then we sang a hymn, privately, there in the police station: "God Bless America."

As the sky turned golden, we walked to the Capitol steps. There were microphones and a podium, and the leadership did return to the Capitol and said a few words backed by hundreds of members arrayed behind them.

Then, unplanned, as the press conference came to a close, someone from the back of the crowd of members on the steps started to sing. Truth be told, I thought at the time I recognized the voice of John Boehner, who was not in leadership at the time. We had just sung the hymn together privately at the police station, and it had affected all of us more than any words.

God Bless America.

I was standing in the front row next to Maxine Waters, with whom I disagreed on just about every policy issue. But it didn't matter. The American people needed to know that we were strong, we were united, and we would endure.

As the sun set, members of the House and Senate talked quietly with each other and then dispersed for the evening.

I still had one more thing to do that night.

Several months before, there had been a discipline and oversight problem at the Page School. At that time, about sixty high school juniors spent the entire year going to school at the Library of Congress and working during the day in and around the House chamber. There were three members of the board of the Page School. The Clerk of the House was one. By tradition, two other members were chosen by the Speaker: one usually a man and the other a woman.

Speaker Hastert had called me to his office to ask for my help. He wanted John Shimkus, a West Point graduate, Army Ranger, and father of three boys, and me to take on that responsibility. Denny knew that Jay and I had been foster parents, had two generally well-behaved young children, and that I had run the child welfare and juvenile justice systems in New Mexico before being elected. We weren't looking for more to do, but John and I understood. We were Academy grads and could probably correct some problems.

The Page School was a great educational experience for the students, but sixteen-year-olds need some clear boundaries and adults who are consistent about consequences when lines are crossed.

Shortly after John and I were put on the board, we went to talk to the pages. We drew straws and I got to be the good cop, telling them how privileged they were to have this opportunity of a lifetime and to take advantage of every day. John played the bad cop and told them about a few important rules like no drinking, no drugs, and being in the dorm by curfew. He made clear that if anyone broke those rules, they would be packed up and sent home before the sun set. A few weeks later, when a student decided to violate one of those rules, we did what we said we were going to do and sent him home. Things settled down at the Page School.

On September 11, I left the Capitol steps and walked to the Page School dormitory. It was dark by the time I arrived. While they weren't expecting me, the staff gathered the students, in pajamas and T-shirts and flip-flops, and we talked. Some hugged pillows or stuffed animals for comfort, others looked at me with a fearful intensity, hoping that I would have answers to their many questions.

I asked each of them if they had been able to talk to their families. They all had.

We talked for a while. I acknowledged that I didn't have much more information than they had seen on television throughout the day.

As the conversation wound down, a boy who had been quiet looked at me intensely and asked, "What should we do now?"

He wanted to know what role they should play, as young citizens, as pages in the House of Representatives.

"We'll go on," I said. "Tomorrow morning you will get up early and go to school. And then, at 10 a.m., when the House opens, I'd like all of you to come to the House floor. I'll be there, and a lot of other members will too, I think. We'll listen to the prayer, and then we'll say the Pledge of Allegiance together. And the business of the country will go on. And you'll be part of it."

That seemed to be enough. A ritual to hold on to. Something to do that would give them meaning and connection to each other and to the country. Doing what we could, where we were, with what we had.

The House did gather the following morning, and Father Coughlin said a prayer and we said the Pledge of Allegiance together.

I started trying to go to the Pentagon, at the request of Kip Nicely and because I knew that New Mexico's disaster recovery team had been mobilized and brought in to help. At first there was resistance, but on Thursday, a group of us went over. Kip, dressed in his Navy whites, met me as I got off the bus. You could still smell the smoke rising from the black scar on the side of the building where the Navy staff was located.

I let Kip talk about his friends, about what happened that day, about the recovery efforts. As we watched, a young Soldier walked out of the remains of the building carrying a Marine Corps flag. I'll always remember looking at him emerge from the blackened rubble with that red flag and, oddly, behind him where the building still stood, on an upper floor was a piece of standard government furniture. It was a book stand with a large government issue dictionary on it. It was intact, open and seemingly untouched a few feet away from where an inferno had torn the building away.

It was a memory that would linger. How could the whole building be ripped apart and, a few feet away, a book lie undisturbed?

Kip and I went into a tent city that had been set up on the grounds of the Pentagon and chatted with the New Mexicans who were experts in disaster medicine and recovery. I thanked them for coming to help. Like most people in the helping professions, they were honored to be there.

I had no idea that my classmate, Dave Goldfein, with whom I would serve in the Pentagon almost twenty years later, was running the tent city that day.

After I returned to the Capitol on Thursday afternoon, we got a call from the Department of Energy. John Gordon was a retired Air Force general who was serving as Undersecretary of Energy. He and I had worked together on the National Security Council staff for George H. W. Bush in the late 1980s. DOE was sending some equipment from New Mexico to New York. The crew would stay overnight at Andrews Air Force Base and go back empty to New Mexico on Friday. If anyone wanted to go, there was space.

There would be a Mass at the National Cathedral in Washington on Friday. But there would be a gathering in Albuquerque that Friday night too. I decided it was more important to be with my constituents in New Mexico than to be one of 435 members of Congress in Washington.

Jim Chavez drove me to Andrews early Friday morning, and I headed west.

I had arranged to meet with those responsible for security at Albuquerque airport. While they reassured me about their plans, we all knew things would change completely for air travel going forward.

I visited several units at Kirtland Air Force Base, including the border patrol aviation unit and leaders at Sandia National Labs. They had already begun modifying threat assessments and developing recommendations for improvements. I would work with them over the next few years to see some of those put in place.

It was early evening when we arrived at the parking lot of the Coronado Mall, where a stage had been erected and several thousand people had gathered for a memorial service.

My family was there to meet me, to be with me, and with our community that was hurting. Because the television had been kept off, our four-year-old daughter had been somewhat protected from the horror of events that week. But she also was much more attuned to people's emotions than her brother.

What she saw, riding on my left hip with my left arm around her, was people approaching me distraught, crying. And I did the best I could to comfort them. But she knew something was terribly wrong.

The sky was darkening when speeches began. City councilors, county commissioners, and other local leaders began to take their turns at the microphone. As the speeches went on, a few people started to rile the crowd and talk about revenge, and they weren't limiting themselves to foreign adversaries or terrorist groups. I knew that Muslim Americans were afraid that they would be blamed or the targets of angry crowds, and, as the speeches went on, I started to become concerned.

As my time to speak came closer, I chose to do something I rarely did. As the speaker before me continued to talk of revenge and raise the level of anger among the people before us, I pulled a business card from my pocket and hastily wrote a few words on the back.

Now it was my turn to approach the microphone. The people gathered that night where angry at what happened to America. They wanted to strike back, and I understood that. But striking back didn't mean punishing the innocent in our community for the sins of others. I knew I had to do something to take down the temperature of the crowd.

I paused at the microphone, glanced at my card, and quietly said, "Would you please join me in prayer."

It wasn't a long prayer and wasn't particularly eloquent, though it was heartfelt. And, at the end of a minute or so with our heads collectively bowed, when I said, "Amen" and walked to my seat, the anger had dissipated, and the crowd turned a corner.

I almost never prayed in public. It seemed to me that too many public figures wanted to be seen to pray. Too often it was insincere. Besides, I had grown up in a small New England town where you didn't discuss faith outside of church—a habit that befuddled one of my evangelical Air Force Academy friends who sought to educate me about the great commission to spread the faith.

But, at that moment, it seemed like the right thing to do with people I cared about who were torn between deep sorrow and roiling anger. We are a people of many faiths, but we are a faithful people. Even as a public leader in a secular government, there are moments when a call to faith is appropriate.

In the months and years that followed, I would spend a great deal of my time as a member of Congress on matters related to national security. But in the immediate aftermath of 9/11, when people were stunned and hurting, the most important things I did as a leader were not legislative initiatives or one-minute speeches on the House floor so much as quietly tending to people and their needs. Being a calm and steady presence for others and, sometimes, doing what needs to be done even if it isn't in the job description, in small ways, behind the scenes, is what leaders must do.

Do what you can,
From where you are,
With what you've got.

Chapter 13

Answer the Call

Heather

The president's office at South Dakota School of Mines is in the same building as its geology museum. The office is modest, befitting a public university that started as a mining school in the late 1800s. From the first day I came on campus in 2013, I felt at home there. There is a rugged independence with a dash of neighborliness in the people of western South Dakota—much as I have found in other parts of the American West. And, for the first time in over a decade, people engaged me not as an elected official but as a university president and community member. I liked that. As I explained to friends, "It's nice to be in a place where people wave at you with all five fingers."

To be sure, winter months were harsh in the Black Hills, but as Senator John Thune once told me, "The cold keeps the riffraff out."

Senator Thune was right. The students at Mines—all engineers and scientists—are certainly not riffraff. There are no easy degrees, and the students are smart and hardworking. There are also no coddled, privileged kids, and I liked it that way. The South Dakota School of Mines and Technology is the best engineering school you've never heard of. It's a school that educates students to a high standard, and industry grabs every graduate it produces. Consistently, over 95 percent of graduates were hired into jobs at an average starting salary of more than $65,000 a year.

Living in South Dakota had its perks, too. I took up fly fishing, finished my instrument rating in a little Cessna 152 that I kept out at the Rapid City airport, and enjoyed hiking and biking in the Black Hills with friends during the gorgeous summers. We had a 5,000-square-foot house to entertain guests and a big room for student game nights. I had a nice car to drive, courtesy of a local dealer who supported the university. Deer and raccoons regularly visited the backyard, much to the consternation of our stupid little dog.

This was all on top of the joys of campus life: concerts, football games, community service projects with wonderful students. And the university system in South Dakota was well run. With only one Board of Regents over the whole system, there wasn't a lot of micromanagement of campus presidents. Mines may not have been the flagship school of South Dakota, but we set out to be the best. I was accountable for results and kept the central office informed, but they didn't tell the campus how to do things. It's not always that way in higher education, so I considered myself to be blessed.

All to say: I had a great job in a beautiful part of the country. I worked hard, enjoyed the company of great people on our leadership team, and woke up each morning excited to get back to work.

Dave

After the 2016 election results were finalized, I asked my special assistant, Sam Neill, to research my role as a service/joint chief during a presidential transition—my first in this capacity—to ensure I was prepared for the task ahead, which was leading the Air Force through a major leadership change while maintaining focus on active combat operations ongoing in the Middle East.

Sam did his usual great work. He put together a few papers highlighting how past administrations formed transition teams to address key issues and fill leadership roles through Senate-confirmed nominations and lower-level appointments. Much was uncertain, but one thing was clear: a new secretary of each service would be key selections once the Secretary of Defense was chosen.

Not long into the transition, I heard that my old boss, General Jim Mattis, USMC, retired, was nominated for Secretary of Defense. I was excited to have the opportunity to once again work for this gifted leader. He was the gold standard for providing clear, concise, and thoughtful command guidance and then trusting his subordinate commanders to figure out how to deliver together. At every gathering, he reminded us he expected "vicious harmony" among his component commanders. It remains one of my best experiences working together with joint teammates.

I had a call with General Mattis, during which he asked me whether I knew Dr. Heather Wilson. He initially didn't share the reason for the question. "Yes, sir, I do," I answered.

He wanted to know more. "Tell me about her."

"Well, sir, we started as classmates at the Air Force Academy in 1978 but traveled in different circles, so we didn't know each other that well." I admitted I was in the part of the class "that made her part the top half."

He chuckled. "I can absolutely relate to that, Dave."

Heather and I did end up crossing paths after the Academy. I shared the story about the last time we saw each other, when I was the wing commander at Holloman

Air Force Base in southern New Mexico and she was a congresswoman from the same state.

"I traveled to Albuquerque to meet her and introduce myself," I said. "When I arrived at her office, there was a long line of people who had the same idea. I waited in line for about an hour talking to the various groups there who wanted something from her, but I didn't hear a single story from anyone there to offer her anything. Such is the nature of our elected representatives."

He listened intently. "When I walked in, Heather met me at the door and gave me a strong handshake. She looked me straight in the eye and said, 'Hello, Dave. So good to see you again! Thanks for taking the time to come see me.'" Here I was, I remember thinking, a mere colonel, and yet she was the one thanking me?

I forgot exactly what we talked about, I told General Mattis, but what I remember is how she made me feel. It was like I was the most important person in the world for the entire time I was in her office. She took notes on a small notepad throughout our discussion. She was completely engaged and genuinely wanted to know about Holloman AFB, our Airmen, the Air Force, and how she could help. She cared. After leaving her office, I saw a few of the folks I met earlier while in line. We swapped stories, and it was clear my experience was not unique. Heather treated everyone with this same respect and focus. What's more, within a week, I received a thank you card expressing her appreciation for my leadership and wishing my family and our Airmen well.

(As a brief aside, having not spent any real time with members of Congress at that point in my career, this experience recharged me about our elected leadership.

The citizens of New Mexico were lucky to have such a caring leader. I was proud of my classmate.)

After sharing this with General Mattis, he asked if I thought she would be an effective Secretary of the Air Force.

"Sir, I don't think you are going to find a better person to lead our Air Force during your time as Secretary," I told him. "I would absolutely love the chance to work with and for her."

He was convinced and asked my help to get Heather to say yes. I told him I'd do my best.

Heather

On a sunny day in November 2016, my phone rang. It was a tad bit cold, and I was in my office at South Dakota Mines looking out the window toward the old gymnasium where the choirs and orchestra practiced.

The voice at the end of the line was grizzled and distinctive. I recognized it immediately, having listened to it on the news.

"This is Jim Mattis."

The election result surprised everyone, including me. Hillary Clinton lost to Donald Trump. He would be the next President of the United States. By this point, the President-elect had announced his intention to nominate retired Marine Corps general Jim Mattis as Secretary of Defense.

"You don't know anything about this, but there's a lot of people who think very highly of you, and I want to talk to you about becoming Secretary of the Air Force."

I paused for a long moment to let the shock sink in. He was direct and clear. I liked that. Then I said, "Sir, you do know that being a college president is, like, the best job in America, right?"

"Yeah, I know," he said. "I just came from Stanford."

"And I didn't apply for any job," I made clear. "I'm kind of a gal of the west, and we kind of like it out here."

"Yeah, I know," he said. "I grew up on the Columbia River in Washington."

My objections weren't working.

We talked a little more. I told him I'd have to think about this and talk to a few people, including the guy I'm married to, because this sort of thing would be a family decision. General Mattis accepted that—even more, seemed to respect it. I did warn him that my husband, Jay, didn't really like rainy, East Coast cities with a lot of traffic. He accepted that, too. We hung up.

I did talk to other people and got their advice about whether the job was worth doing, what Mattis was like to work for, and what I would be giving up.

General Mattis and I talked a few more times. I told him a quick background check would reveal that I had contributed to one of the President-elect's primary opponents. He didn't care. "I want the best team on the field," he said.

I also told him that he should have his legal beagles look into a Department of Energy Inspector General's investigation, as some people might use it to cause problems with my confirmation. He thanked me for mentioning it and subsequently said the lawyers had reviewed it and didn't think it was worth worrying about.

"At least nobody died in your IG investigation," he commented. It sounded like there was a story there, but I didn't press the point. I did, however, press upon him several other people he should talk to about the job, people who might do well and might want the job more than I did.

I also told him that service secretaries seemed more often than not to be rewards for friends of the President, similar to ambassadors. They cut ribbons and traveled a lot but didn't always do important things. Mattis assured me he wasn't looking for someone to cut ribbons. He needed someone who could restore the readiness of the force, pivot toward the Pacific, and improve the procurement process. He said he thought the service needed someone who understood Airmen and that I seemed to have a certain "genetic predisposition" toward the air service. Furthermore, he correctly understood that relationships on Capitol Hill were frayed and that the new Secretary would need to spend a lot of time with Congress. In his view, I was ready for that, too.

I kept thinking about it.

I was almost to the point of politely declining when I talked to two classmates. Bob Otto had just retired as a three-star deputy chief of staff and was one of my closest friends at the Academy. He strongly encouraged me to take the job, if offered, and said he thought someone like me could make a tremendous difference.

Then I called Dave Goldfein, who had recently become Air Force Chief of Staff. We hadn't talked in years and didn't know each other well as cadets, but we had started at the Academy on the same day. I needed to get an early sense of whether Dave and I could build a strong partnership. The conversation went well, and I hung up the phone feeling confident he was someone I could work with.

Dave

Having been nominated as Chief of Staff of the Air Force by President Obama and confirmed by the Senate Armed Services Committee, chaired by John McCain, I understood the importance of navigating the vetting process with sensitivity and adhering to Senate protocol. Lesson number one for anybody nominated, or anybody who knows somebody who is or might be nominated, is don't presume confirmation. With this in mind, I was very careful about following the rules.

I asked the Air Force General Counsel whether I could invite my former USAFA classmate over for a private dinner at the Air House while she was in the process of consideration for the role. After review, they gave me the green light. While the law prohibits taking action that presumes one's confirmation, they informed me, it does not inhibit taking action that properly prepares one for the role. In fact, such action is expected. They advised I share with Heather only what I would otherwise share openly with leaders in industry or academia.

Even though our paths had crossed over the years, Heather and I hadn't seen each other since our meeting ten years earlier at her congressional office in New Mexico. Like then, our conversation over dinner at the Air House was spectacular.

Heather

Dinner at the Air House with Dave was serious yet punctuated by laughter. Dave's wife, Dawn, welcomed me and we caught up. Then, she left the two of us to have dinner and talk about the Air Force.

I had to start by clearing up a memory.

"OK," I said. "When we were cadets, there was a talent show and someone sang a song called, 'Going Ac Pro.' Was that you or your brother?"

"Ac Pro" was the cadet term for being on Academic Probation. The song was the highlight of the talent show and brought the whole audience of cadets to our feet cheering and laughing. I still remembered it more than thirty years later.

"That was me," Dave said.

"That was hilarious," I said.

"Yeah, they asked me to sing it from the staff tower at lunch the next day. The Commandant of Cadets didn't think it was so funny. I did two weekends of tours for that song."

A "tour" was a punishment for cadets that involved marching in a square on a Saturday in dress uniform with a rifle on your shoulder.

I laughed again.

"But it was worth it," Dave said with an impish smile.

Laughter and self-deprecating humor are tools used by gifted leaders. Dave Goldfein had mastered this skill better than anyone I ever met. I quickly started to get the feeling that we not only could work together but that this would likely be fun, too.

We talked about a lot of things important to the Air Force that evening: readiness and procurement, pilot retention and the pace of operations, and relationships with the other services. I had a lot of questions and learned a lot. I left thinking that, even if I didn't become the next Secretary of the Air Force, the country was in good hands.

Dave

By the end of dinner, I knew that Heather was the same smart, focused, and engaged leader I remembered from my visit to Albuquerque. Like then, she wanted to know about the challenges we faced as a service, where we needed congressional support, the state of our readiness, and the welfare of our Airmen and their families. The dinner clarified these questions, but it mostly served as an opportunity for us to individually assess how well we might work together. Our instincts were right: We knew we would make a great team, and we did.

Heather

Jay and I talked at length about the Secretary job. Jay is a retired Air Force colonel who spent over thirty years in active, Guard, and Reserve service. He had no strong desire to live in Arlington, Virginia, but we shared common values. There was a deeply personal element for him, too. Neither of our two children served in the military, but one of Jay's former Boy Scouts, Matthew Roland, had gone on to the Air Force Academy and had been killed in Afghanistan. Jay remained close to Matt's father.

Jay has always supported me, and when he reflected quietly in one of our conversations that our country hasn't asked much of us, I knew he was thinking about Matt. We hadn't lost a child like other families. Salary and houses and board positions seemed trivial in comparison.

In early December, Jim Mattis called again. The former Marine Corps recruiter had me figured out.

"Look," he said. "I've talked to a lot of people. You are my first choice and there is a big gap between you and my second choice. I want to know if you will serve, if asked?"

I heard what he said, but deep down I understood what he was really asking: *Heather, who are you?*

There are moments in life when you're confronted with questions that cut to the core of who you are—your identity, values, and purpose. These are not easy questions; they challenge us, unsettle us, and often leave us searching. But they are also the most important questions because they shape the way we live, lead, and leave our mark on the world. And sometimes, you are asked to serve in ways that look different than your best-laid plans and may not be in your short-term interests. In my case, saying yes to General Mattis would mean leaving a job I loved and a place we really felt at home. It would mean resigning from two corporate board positions and a university job I loved and taking a substantial cut in pay.

But in that moment, I knew that if I said no and continued to urge my students to use their gifts in service beyond themselves, then I was a fraud.

"Sir, I will serve, if asked."

Then, I waited. I knew that it was entirely possible that the President-elect might have someone else in mind for the job. If that was the case, I'd be very happy to stay in South Dakota.

A few weeks later, as we were leaving the Detroit airport in a rental car headed for Christmas with Jay's relatives, my phone rang. It was Jim Mattis.

"I have been authorized by the President-elect to tell you that he would like to nominate you to become the next Secretary of the Air Force."

I looked at Jay.

"Sir, I will serve." No turning back now.

Then he told me to brace for an ungodly amount of paperwork.

As it turned out, I really enjoyed being Secretary of the Air Force. Secretary Mattis was one of the best leaders I ever worked for. He gave clear guidance on objectives and had very high standards, but he didn't micromanage. He cared deeply about the security of the country and inspired us all with his dedication to service above self. He made corrections in private and complimented in public. He even translated his Marine Corps jargon for those of us left blinking quizzically at some of it, usually to laughter.

Dave Goldfein was also a wonderful partner. Life has a way of working out, and on the other side of it I was able to return to higher education in the west with even more skills and perspective than I had when I was at South Dakota Mines.

All things I didn't know at the time. And all things I wouldn't have learned if I hadn't answered the call.

Answer the call.

Chapter 14

We Are Our Stories

Heather

"Is the Chief available?" I asked, as his military aides popped to their feet, a little surprised, in Dave's outer office.

As Secretary and Chief, our office suites were next to each other on the outside ring of the Pentagon, just upstairs from the Office of the Secretary of Defense. It wasn't unusual for one or the other of us to walk down the hall, unannounced, and poke our head in to talk about something. That informality helped build our working relationship, and the staff gradually became accustomed to it.

The lieutenant colonel at Dave's front desk told me the Chief was in and walked over to his reinforced door to alert him, and then stepped aside to let me pass.

Most Pentagon leaders have artwork in their offices that is meaningful to them. The Pentagon has a large art collection, and you can ask for pieces to be hung in your office. My office displayed paintings and artifacts celebrating scientific and technological advances of the air service, like the SR-71 and the first crew that refueled an aircraft in flight. I also had a painting of a little single-engine high wing observer plane backed by beautiful clouds because it reminded me of just how small we are in the big scheme of things, and it looked like my little Cessna I had left behind in South Dakota.

Dave was also one for meaningful displays in his office.

I couldn't help but notice an arresting photograph of an Airman in combat gear holding a child that hung prominently behind his desk. The man in the photograph was somewhat disheveled, but his eyes were piercing and intensely bright.

"Who's that?" I asked.

"We need to talk about that," he said. "There's a package on its way to you for the Medal of Honor."

Dave Goldfein was the first one to tell me the story of John Chapman.

Dave

My longest serving executive officer was Colonel Wolfe Davidson. Wolfe is really his middle name, growing up in a family that named the kids after their favorite animals. It turned out to be appropriate for him because he became a special warfare officer—a combat controller—who served in one of the most elite combat units in the military.

Wolfe not only brought a refreshing balance of humility and professionalism, but he also helped me stay connected to our special operators serving as the ground component of our air-ground battle across Iraq and Afghanistan. As the former commander of our wing responsible for the highest-level of special operations, he brought a credibility with the other services that was essential to success.

At the end of one long day in the office when we were packing up my nightly read file, Wolfe told me the story of Technical Sergeant John Chapman.

"I was John's flight commander at the Special Tactics Squadron," he started. "They don't come any finer. John was not supposed to be on the deployment where this picture was taken, or the mission that would be his last. On 9/11, John was serving in a high visibility staff job, getting a break from the nonstop training trips and deployments on special tactics teams."

Air Force Special Operators are unique among the services as they join Navy SEAL and Army Special Operations teams to bring airpower to the mission. As a result, their training is the equivalent of both the Navy and Army combined to become fully capable members of either team.

"John and his wife, Valerie, had two daughters under five, Madison and Briana. Understandably, the family needed a break from the typical 250+ days a year on the road. The staff job gave him that."

This was the case for so many of our special operations Airmen. They were in great demand even before 9/11, and their families paid the price. Military families have a very special kind of courage. They manage multiple deployments, long work hours, dangers facing their loved ones, and a loneliness that comes from raising children while your spouse is out doing the nation's business.

"When the first special tactics teams deployed in October 2001, we sent John to support the Navy SEALs," Wolfe told me. "By the end of December, we had trained a younger Airman who we planned to replace John."

On Christmas Day that same year, John came into the unit headquarters and pitched an alternate plan. He looked Wolfe in the eye and said, "This SEAL team is the most experienced special operations team deploying, and they will be sent on the most demanding missions in Afghanistan. They need the best combat controller we can provide. I need to deploy with the team."

These were surprising words because John Chapman was a humble man. But he was right. He was the best man for the job, and his team needed him.

Which is why Wolfe changed the deployment orders and sent John on his final combat tour.

Valerie and the girls never complained. They knew this was something John had to do. They also knew, as John did, that he wouldn't forgive himself if he stayed behind.

So, John and the SEALS deployed to Afghanistan.

On March 3, 2002, John and his team were given a reconnaissance mission to go to the top of a mountain known as Takur Ghar, in the southeastern part of the country. Their mission was to establish an overwatch position so John could guide the waves of airpower that would support a ground assault in the valley below.

Two days later, the mountaintop would become known as Roberts Ridge.

As the Army MH-47 helicopter carrying John and the SEALs was settling into the landing zone, gunfire erupted. The aircrew was able to get the aircraft out of the kill zone, but the helicopter immediately descended to a crash-landing several thousand feet below the peak. When they counted heads, the team leader realized one of the SEALs, Petty Officer Neil Roberts, had been thrown from the helicopter during the assault.

John immediately made radio contact with the Air Force AC-130 gunship that was over the landing zone and received the dreadful news that Neil Roberts was on the landing zone and fighting but was surrounded by the enemy.

Leaving one of our own behind is contrary to everything we are trained to do and be. The AC-130 gunship overhead could see al-Qaeda forces making their way up the ridgeline to kill Roberts. They laid down suppressive fire, but someone had to go back to get Neil.

John and the team unanimously agreed to rescue their teammate. John called for another helicopter to take them back to the ridge.

He was the lone Airman paired with the Navy SEAL team as they flew back in to rescue Neil Roberts. The MH-47 Chinook landed in a bowl on the snow-covered mountaintop near Roberts's last known location. As the team exited the ramp of the aircraft into thigh-deep snow, the helicopter quickly departed, and the team came under withering fire from several directions. The most threatening fire came from an elevated position with a rock bunker about thirty meters away up a steep slope.

A drone aircraft overhead captured all the action on an infrared pod, friend and foe. Every living thing on that mountain showed up as a bright warm spot against the deep snow. When bullets were fired, they could be seen as hot flashes in the video.

It would become the best evidence of John's heroic actions that night.

As John departed the aircraft, he quickly identified the fighters in the dominant position and instinctively moved toward the enemy, firing frequently as he climbed through the snow up the hill. The drone recorded John fighting as he made his way ahead of his SEAL teammates toward the rock bunker where several al-Qaeda fighters were also shooting. He destroyed the enemy and eventually made it to the bunker where he could have safely waited for the team. However, upon identifying a second fighting position with a heavy machine gun less than ten meters away, John left the cover of the bunker to engage this second position.

Exposing himself to what had to be done allowed another fighter to shoot John at close range. John returned fire, took out the fighter, then fell in the snow.

The fighting on the mountaintop grew intense as more Taliban fighters made their way to the hill. When the team leader reached him, John was covered in blood and seemingly lifeless. Then another teammate was hit and wounded. The team was taking casualties, and options were dwindling. They fought on in a battle for survival. The team did not get back to John.

The team leader got the rest of the SEAL team off the mountain successfully and safely—an incredible act of leadership for which he would later be awarded the Medal of Honor.

Then something remarkable happened. Shortly after his teammates left, against all odds and expectations, John started moving again. The drone footage showed it clearly. He was not dead. He was critically wounded but would continue to fight for another hour. Video recordings show that during the final sixty minutes of John's life, John engaged with at least another four enemy fighters attacking his position.

Unbeknownst to them, there was a special tactics combat search and rescue team with an Army Ranger quick reaction force dispatched to rescue the battered team.

John heard and saw the helicopter flying in to rescue him and his teammates. He moved to higher ground to engage a fighter who was preparing to shoot a rocket-propelled grenade at the approaching American helicopter. John exposed himself to the enemy again to stop him before he could fire in a last attempt to save the helicopter and its crew. As John continued firing and advancing toward the enemy, he was shot and killed.

John used his remaining strength and sacrificed his own life to save his teammates.

When his body was eventually recovered, it had nine bullet wounds, one of which was determined to have been instantly fatal.

Wolfe recounted this story to me and sat quietly after he finished. He then showed me the video.

Heather

A few days after Dave told me the story of John Chapman, Lisa Disbrow, my acting Undersecretary of the Air Force who had stayed on for the transition, came to see me with Colonel Wolfe Davidson. They gave me a thick folder to review: a recommendation for the Medal of Honor for Technical Sergeant John Chapman. It had been sent to the White House during the previous administration for President Obama's approval, but time ran out. With a new administration and a new President, the recommendation had to start over and go through the entire process again.

As he did with Dave, Wolfe showed me the video from the AC-130 and drone footage from overhead. I watched it several times, bearing in mind what I'd seen as I carefully read through the thick file later that evening at home.

Dave

For every Medal of Honor in history, someone had to swear to the courage and sacrifice of the individual based on an eyewitness statement. In our case, John was alone on that snowy mountain top. Our most compelling evidence was the drone video, rather than a firsthand account. Would a video be enough? Would the Department of Defense reverse its centuries-long requirement for an eyewitness account?

When Heather finished reading the file, we sat down to talk about what we should do next. We agreed that the evidence strongly supported that John Chapman fought alone for an hour on that ridge, that he had taken extraordinary risks to save his comrades, and that his actions were commensurate with others who were recognized for displaying similar valor.

We decided he had earned the Medal of Honor.

I would take the case to the chairman and fellow Joint Chiefs, and Heather would talk to her fellow service secretaries as well as the Undersecretary of Defense. We'd work together to persuade Secretary Mattis, who would be the final decision-maker before sending a recommendation to the White House.

Heather

Having been assured of support from the acting Secretary of the Navy, I made an appointment to see Bob Work, Acting Deputy of Defense, and sent the file with my draft recommendation to him. Bob was a former Marine Corps officer. He was a holdover from the previous administration and wouldn't stay much longer. He knew the case and some of the potential objections from elements of the Navy. I found him to be honest and effective. I told him I knew there were some people in the Navy special warfare community discouraging the Pentagon from moving the nomination forward.

"I've gone through all the records, Bob, and I've looked at the film," I said. "I believe John Chapman fought on alone on that ridge and saved his comrades at the cost of his own life."

I told Bob that I knew there would be some unhappy people in the SEAL community for recommending this award, but, while they are a little bit quieter about it, there were Airmen who felt just as strongly that John deserved to be recognized.

After asking a few questions, Bob Work looked at me and said, "Well, the best thing to do, then, is to do right by 'Chappie' and his family. I'll support this."

Dave

The chairman of the Joint Chiefs of Staff was General Joe Dunford. "Fighting Joe" was, like Bob Work, a Marine and previously served as the commander in Afghanistan where I was his lead Airman. We had formed a relationship of trust and mutual respect during some tough years fighting against al-Qaeda and the Taliban. As friendships go, those born in combat are the deepest and most lasting.

After a session in the tank with my fellow Joint Chiefs, I asked the chairman if I could walk back with him to his office to discuss a sensitive issue. I explained the situation with John Chapman and asked if he would be willing to allow my executive officer, Wolfe Davidson, the chance to sit down with him to discuss the case in greater detail. I knew Joe would be fair, having done the same for several Soldiers and Marines under his command who had earned the Medal of Honor. I also specifically told him I would not attend. I wanted to know whether he thought it merited advancing the package to the Secretary of Defense and eventually the President for approval when we would be relying so heavily on video evidence for the first time in history.

He agreed to the plan, and shortly thereafter Wolfe walked Chairman Dunford through the firefight and the video. Joe spent almost two hours with him, and, by the time they were done, the chairman was convinced. There was no denying what the video evidence revealed about the final hours of John Chapman's life.

Chairman Dunford called me after the meeting. "Dave," he started, "Thanks for sending Wolfe down to see me. He is quite an officer." I wasn't sure how this was going to go. "I looked at the video," he continued, "and I am convinced John Chapman deserves the Medal of Honor." It was as much a relief as it was a wave of gratitude and joy. "Wolfe was able to answer all my questions," he said. "I'll take it from here."

Once I knew Joe Dunford believed the evidence and was on board with what we wanted to do, it was time to tell my Navy counterpart, Admiral John Richardson. It was a joint Navy–Air Force operation, and we wanted to be completely aligned. John is among the most thoughtful and sincere officers I have ever worked with, and I knew he would do the right thing. I asked him if he would similarly meet with Wolfe and go through the story. He agreed.

Wolfe walked my friend and fellow chief through the case. When he was done, John called me back.

"Fingers, John Chapman deserves the Medal of Honor. I'll work my side of the issue."

John Chapman was a hero. And John Richardson is a good man.

Chairman Dunford gathered all the affected parties to express his full support of the package. Our next step was to bring it to Secretary Mattis.

Heather

The recommendation for a Medal of Honor comes from the Service Secretary to the Secretary of Defense. But Dave and I knew that Secretary Mattis would ask others around him for their advice. When we were sure that those Secretary Mattis would likely turn to had been briefed and were on board, I sent my written recommendation forward.

It was a matter of time before the signed package from the Secretary of Defense was on its way to the White House for the President's approval. Chairman Dunford and Secretary Mattis also approved a package submitted by Admiral Richardson for the same award to the SEAL senior troop chief, Britt Slabinski, for getting the rest of his team off that mountain despite all the odds stacked against them.

It took a few months, but we eventually got word that the Medal of Honor had been approved for John Chapman. At that point, we celebrated the successful outcome then shifted our focus to determining how best to tell his story for it to be meaningful to the Chapman family and the United States Air Force.

Too often, leaders think that the bottom line is all that matters. At quarterly meetings and in annual reports, they dwell on spreadsheets and PowerPoint charts. Data has its place. But as a leader, you need to tell stories that illustrate the best of your organization to your members.

We are our stories. Every person you lead has a story. Sometimes it's a story of exceptional service and sacrifice, like John Chapman's. As a leader, it is your job to uncover stories and to cultivate an environment for stories to emerge. Stories inspire and uplift. They also bring people along and show them what "right" looks like.

Dave

We worked for months planning how we would celebrate and honor John Chapman and his family. We wanted to get this one right.

The night before the ceremony, Dawn and I invited Heather and her husband, Jay, and other Air Force leaders along with the entire Chapman family over for dinner at the Air House.

The Air House is the residence of the Air Force Chief of Staff. It stands on a hill overlooking the field where the Wright brothers first demonstrated their flying machine to the Department of the Army in 1908. It overlooks the Potomac River to the Jefferson Memorial and, beyond that, the celebrated monument of our nation's first President.

After a lively dinner, we assembled in the living room in front of a beautiful glass window with a view of Washington, DC, and Arlington Cemetery, where so many heroes lay. The Air Force band provided musical entertainment, including the closing song "God Bless the USA" by Lee Greenwood.

When the band arrived at the chorus line—"and I'll gladly stand up next to you, and defend her still today"—Valerie and her daughters rose from their chairs and began singing with the band. Everyone in the Air House stood up and joined in. We locked arms in a circle and swayed, singing as the sun set over Washington, all of us seeing the same image in our minds of John Chapman racing up Robert's Ridge to save his teammates.

We sang, arm in arm, through tears.

Heather

The next day, the big day, started with the presentation ceremony at the White House. President Trump recounted the story of John's actions on March 4, 2002, that went above and beyond the call of duty. He called Valerie up on stage. It had been fifteen years since John's passing. She had been so very patiently waiting for this moment. When the President handed her the plaque and medal, Valerie let out a whoop and held it in the air to a standing ovation. The President smiled and applauded as we all took in the moment.

It had been quite a journey to get here. John's teammates knew the day after the battle that John's final actions were not accounted for in his initial Air Force Cross citation. John's leaders and teammates spent years working to account for all his actions, and they had pledged to John's family that they would make sure his story was told. For our part, we saw through two administrations and built consensus across the senior levels of the US government.

The White House event was certainly one to remember. But to me, the more special ceremony took place that afternoon in the Pentagon. Hundreds of John Chapman's comrades gathered to celebrate his story and put his name in the Hall of Heroes, a passage in the Pentagon dedicated to honor those who have earned our nation's highest award for bravery, where more than 3,500 names are displayed on its walls.

Dave brought the picture that had been hanging in his office, on loan from Wolfe Davidson.

Dave

I decided to bring with me to this part of the day's program the framed photo that Wolfe loaned me of John Chapman in Afghanistan. The same one that had hung in my office behind my desk for the previous four years. It showed John holding a young Afghan girl in his lap who happened to be the same age as his youngest daughter, Brianna, at the time. The photo was taken just a few weeks before his last mission, and the story behind it felt to me a perfect and fitting illustration of the man we gathered to remember.

Local Afghan operators were clearing a house to make room for the team that would scale Roberts Ridge. As they were removing the family, John stopped them and told his Afghan counterparts to keep the family where they were since there was heat in only one room and the kids were cold. When they disregarded his request, he picked up the little girl and held her on his lap. If they wanted to move the family, they would have to deal with him.

As I showed the picture to guests, I shared its unique story but also how it represents the incredible men and women I was privileged to serve as their Chief. People who one day are holding a little girl in their lap to keep her family in the one warm room in their home, who on the next day are unleashing fire and fury on the enemy. What an honor it has been to lead the John Chapmans of the Air Force.

Heather

It's hard to follow Dave at any event because he's so good at telling stories. But I did my best that day to tell the story of John Chapman and the sacrifices he made for our nation before a Pentagon auditorium filled with John's family, his friends, and a sea of red berets.

Dave

The following morning at the Air Force Memorial, we gathered to unveil John's name on the Medal of Honor memorial wall. Valerie moved the crowd to tears as she called up the photographer to take a picture from where she was standing. She wanted to remember that sea of faces and young special operations warriors who came to honor John.

At the end of the ceremony, everyone present assembled for memorial push-ups, a special tactics tradition. I got down into my front-leaning rest position as we prepared to execute on the customary command: "The next exercise will be memorial push-ups. Position of exercise: move!" Next to me were two wounded warriors who had both lost limbs in Afghanistan. They intertwined what was left of their arms and legs to complete their push-ups as one body. The wall of names, Valerie standing before the crowd, wounded warriors taking to the floor—all images as impossible to forget as the man we were assembled to remember.

Such moments are singular, and yet each one inevitably surfaces a similar reflection. Once again, I asked myself, "Am I worthy to lead men and women like this?"

Heather

When the push-ups were done, the crowd lingered, catching up with friends and sharing stories. We'd spent the past three days honoring a fallen Airman in a way that I will never forget.

There was still one more thing I wanted to do.

I noticed that the military aircraft used by the Army for senior officers and civilians had tail numbers and stories. Tail number 1945 was nicknamed "The Omaha Beach," for example, and inside the main cabin there was a plaque dedicated to the operation and small cylinder of sand from that shore. I thought it was a nice touch. It caused people to pause just for a moment and reflect on the heritage of the United States Army.

Before I left the Pentagon, we changed the tail number for an aircraft that flies special missions. Tail number 2002, after the year he died, is the aircraft named for John Chapman. And inside the cabin is a plaque with his story and a picture of John in Afghanistan holding a little girl in his lap.

Studying history makes us smart, but heritage makes us proud. Leaders need to weave the stories of those who reflect the best of their organizations into the fabric of their cultures. The effect will inspire others who will take the organization forward.

We are our stories.

Epilogue

Being Airmen and leading large organizations has taught us a lot along the way. We always found it less painful to learn from other people's mistakes rather than our own, so we wrote this book for the next generation of leaders who want to be good at what they do. We hope you learned something from it.

We also wrote it for the next generation of our own families: our children and grandchildren, whom we enjoy watching make their own pathways to satisfying lives in the service of something bigger than themselves.

As you probably gathered from our stories, our families matter a lot to us—and they matter a lot to those whom you are privileged to lead. While we want our people to bring their best selves to the mission every day, most people work to put food on the table and to make a better life for those whom they love. If you keep that in mind, providing flexibility when needed for people to care for their families, you will have talented, caring people who want to work with you even if the hours are long and the pay is better someplace else.

In our case, we also have family to thank for this book. Rachel Hone, Heather's daughter-in-law, serves the country in her day job and has been a great editor for us. In fact, she has turned lemons into lemonade so well that she earned her call sign: Lemonade.

And, of course, our spouses, Jay and Dawn, who are leaders and servants themselves and have both supported us along the way. We both married well, and that has made all the difference.

Made in the USA
Monee, IL
30 September 2025

30414806R00105